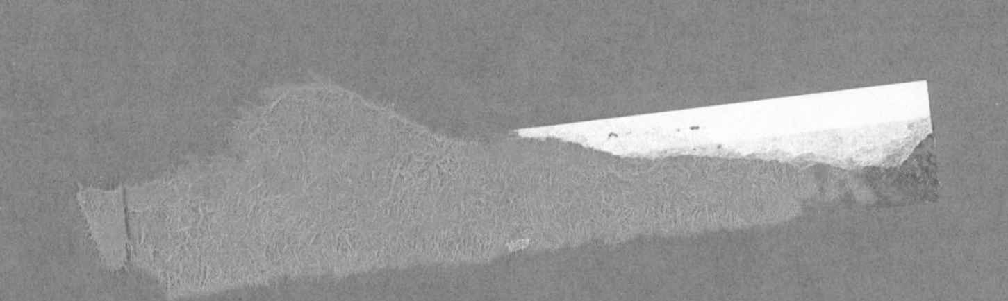

A publication in
The NORC Series in Social Research
National Opinion Research Center
Norman M. Bradburn, Director

Total Survey Error

❋ ❋ ❋ ❋ ❋ ❋ ❋ ❋ ❋ ❋ ❋ ❋ ❋ ❋ ❋ ❋ ❋

Applications to Improve Health Surveys

Ronald Andersen
Judith Kasper
Martin R. Frankel
and Associates

✲ ✲ ✲ ✲ ✲ ✲ ✲ ✲ ✲ ✲ ✲ ✲ ✲ ✲ ✲ ✲ ✲

Foreword by Odin W. Anderson

Total Survey Error

✲ ✲ ✲ ✲ ✲ ✲ ✲ ✲ ✲ ✲ ✲ ✲ ✲ ✲ ✲ ✲

Jossey-Bass Publishers
San Francisco • Washington • London • 1979

TOTAL SURVEY ERROR
Applications to Improve Health Surveys
by Ronald Andersen, Judith Kasper, Martin R. Frankel, and Associates

433 California Street
San Francisco, California 94104
&
Jossey-Bass Limited
28 Banner Street
London EC1Y 8QE

Library of Congress Cataloging in Publication Data

Andersen, Ronald.
Total survey error.

(NORC series in social research) (Jossey-Bass social and behavioral science)
Bibliography: p.
Includes index.
1. Health surveys—Statistical methods.
2. Errors, Theory of. 3. Health surveys—United States—Statistical methods. I. Kasper, Judith, joint author. II. Frankel, Martin R., joint author. III. Title. IV. Series: National Opinion Research Center. NORC series in social research.
RA409.A75 614.4'2'0723 79-88104
ISBN O-87589-409-7

Manufactured in the United States of America

JACKET DESIGN BY WILLI BAUM

FIRST EDITION

Code 7922

The Jossey-Bass Social and
Behavioral Science Series

Foreword

This book is the fulfillment of a wish I have had for many years but did not have the capacity or competence to write myself. Those interested in social survey methodology in relationship to health services use and expenditures now have a source to which they can turn. The pioneer in social surveys of use and expenditures is the one sponsored by the Committee on the Costs of Medical Care.* That survey was conducted at a time when sampling methodology had not been developed as it was in the early fifties, when I was associated with the first such household sur-

*I. S. Falk, M. C. Klem, and N. Sinai, *The Incidence of Illness and the Receipt and Costs of Medical Care Among Representative Families: Experiences in Twelve Consecutive Months during 1928–31,* Publication on the Cost of Medical Care, Report No. 26 (Chicago: University of Chicago Press, 1933).

vey since 1931. Nor were questionnaire construction and interviewing techniques as refined.

It would seem that the beginning of refinement in social survey methodology for use and expenditures for health services is embodied in the survey conducted in 1953 by the Health Information Foundation, New York City, when I was its Research Director, and by the National Opinion Research Center, University of Chicago, with Jacob J. Feldman as my methodological collaborator. Indeed, the name of Feldman threads through the subsequent survey in 1958, when he was again a collaborator, through the surveys in 1963 and 1970, when he was a consultant.* Feldman is currently associate director for analysis, National Center for Health Statistics, Department of Health, Education, and Welfare.

It was my good fortune to be unwitting custodian of a legacy in that I was a research assistant to one of the authors of the original survey sponsored by the Committee on the Costs of Medical Care, Nathan Sinai, and later to encounter Ronald Andersen when I was seeking a project director for the 1963 survey. It is apparent that Andersen and his colleagues have mastered both content and methodology, and this volume is thus a methodological culmination as well as a data culmination up to the present time. The social surveys listed contained brief methodological sections to validate the content, but this book represents an exhaustive statement and critique on the methodological problems in household surveys as to accuracy and validity of reporting as compared with actual records in hospitals, physicians' offices, and third-party payers. It is apparent that household informants are accurate enough for the comprehension of the macro-aspects of health services use and

*The full references are: O. W. Anderson and J. J. Feldman, *Family Medical Costs and Voluntary Health Insurance* (New York: McGraw-Hill, 1956); O. W. Anderson, P. Collette, and J. J. Feldman, *Changes in Family Medical Care Expenditures and Voluntary Health Insurance: A Five-Year Resurvey* (Cambridge, Mass: Harvard University Press, 1963); R. Andersen and O. W. Anderson, *A Decade of Health Services: Social Survey Trends in Use and Expenditures* (Chicago: University of Chicago Press, 1967); and R. Andersen, J. Lion, and O. W. Anderson, *Two Decades of Health Services: Social Survey Trends in Use and Expenditure* (Cambridge, Mass: Ballinger, 1976).

expenditures seen from a national policy perspective. This revelation in itself will therefore not necessitate the great expense of validating household responses with actual records. The degree of accuracy necessary for other applications will have to be judged on their merits.

Chicago, Illinois
July 1979

Odin W. Anderson
Director, Center for Health Administration Studies
University of Chicago

Preface

Total Survey Error explores the validity and reliability of data collected in a large national survey of health care use and expenditures. As researchers and policy makers come to depend on surveys as primary sources of data, it becomes increasingly important to assess the accuracy of the information being collected. Mauldin and Marks, two census researchers, stated the problem over twenty-five years ago: "In common with other organizations the Census Bureau is becoming increasingly aware of the gap between getting *an* answer and getting a correct answer" (1950, p. 657).

Since the development of survey sampling techniques in the late 1930s, survey samplers have recognized that sampling variance represents only one of the many possible sources of total survey error. This early awareness is apparent in the first two texts devoted to survey sampling (Deming, 1950; Hansen, Hurwitz, and Madow, 1953) and in numerous articles published during the 1940s and early 1950s (Bancroft and Welch, 1946; Cornfield, 1942; Eckler and Pritzker, 1951; Feldman, Hyman, and Hart, 1951; Palmer, 1943; Stock and Hochstim, 1951).

Despite this early recognition, measurement and estimation of total survey error sources, other than sampling variance, were rarely attempted. In particular, discussion of error under the general heading of nonsampling bias was typically relegated to the technical appendices of survey reports. In most cases this discussion consisted of statements acknowledging potential differences between reported sampling errors and total survey error. The reader was warned that the sampling errors the researcher reported should not be interpreted as taking into account all sources of error in the reported survey estimates.

This study describes one effort to measure the impact of bias (from nonsampling sources) and variable error on health survey estimates. This attempt to provide an operational, quantifiable definition of the concept of total survey error should be interesting and helpful to the general survey research community and especially relevant to those concerned with social policy and health services.

The magnitude of errors associated with social survey estimates of behavior for population groups of varying age, income, race, residence, and health status is of particular concern in public policy research and evaluation. Definitions of social problems, as well as subsequent planning and implementation of corrective programs, are often based on apparent behavioral differences among such groups. Those who plan, study, and carry out social policy should find relevant those sections of this volume that examine whether differences among groups may be larger (or smaller) than they first appear when error variance by group is taken into account.

The estimates examined are specific to health services delivery—amount of health services people use, expenditures for these services, and methods of paying for them. Consequently, in addition to issues of interest to survey research and social policy audiences, we hope to address the reliability and validity of several indicators of the health services system performance. These issues concern health administrators, medical care providers, and health services researchers.

The 1970 Survey

The 1970 Center for Health Administration Studies-National Opinion Research Center of the University of Chicago (CHAS-NORC) survey was the fourth in a series of national health surveys collecting information on health services use and expenditures in the United States. The survey was supported by the National Center for Health Services Research under contracts HSM 110-70-392 and HRA 106-74-24. In early 1971, 3,765 families plus some additional aged individuals (11,619 persons in all) were interviewed in their homes. One or more persons in each family provided information on health services use and related costs during 1970, insurance coverage, attitudes toward providers, and health beliefs. The sample constituted an area probability sample of the noninstitutionalized population of the United States. It was designed to overrepresent the inner-city poor, aged, and residents of rural areas in order to permit more detailed analyses of these groups. Weights were developed to correct for the oversampling of these groups and to allow estimates to be made for the total noninstitutionalized U.S. population. (Chapter Nine discusses these and alternative weighting procedures.) The weighted response rate to the survey was 82 percent. (For details of the sample design, see the methodology section of Andersen, Lion, and Anderson, 1976.) In addition to data provided by the sample families, a second body of data was collected from family physicians, clinics, hospitals, insuring organizations, and employers about the families' medical care and health insurance for the survey year.

This body of data, referred to as the "verification," was unique to the 1970 survey and had two purposes: first, to determine if the reported care was in fact provided during the survey year and, second, to elicit more precise information than the families were likely to give on diagnoses, costs, kinds of treatment, and sources of payment for services. Approximately 10 percent of the families in the social survey refused to sign the permission forms needed to obtain information from doctors and hospitals. Other families signed permission forms but failed to provide enough information to enable us to contact their providers. In addition, some providers refused to cooperate with the survey. However, verifying data were obtained for over 90 percent of the hospital admissions and for two thirds of the physician visits sample families reported. (Persons who refused to allow verification and providers who did not cooperate are examined as problem respondents in Chapter Six.)

Another difficulty in the verification process was the problem of "false negatives." Although reports of use or expenditures were verified when possible, it was impossible systematically to verify reports of nonuse. Therefore we were more likely to discover persons who reported experiences they did not have (overreporting) than people who failed to report experiences that occurred (underreporting).

The existence of verification data enables us to evaluate nonsampling error. This evaluation is based on the assumption that what doctors and hospitals report is closer to reality than what survey respondents report. Of course, there is also error in the provider reports (Marquis, Marquis, and Newhouse, 1976). However, as other authors have noted, an operational measure of the "true answer" must exist if we are to assess individual response error (Ferber, 1966; Hansen and others, 1951). The importance some researchers place on validity criteria is indicated by Madow's (1965) suggestion that virtually every behavioral science survey should be accompanied by a validation study to serve as a basis for detecting nonsampling errors.

Organization and Major Findings

Total Survey Error is divided into four parts: overview, estimates of survey error, alternative approaches to improve survey estimates, and summary and implications. The first part discusses the model of total survey error used throughout the book—a model drawn from Leslie Kish's classification of sources of bias and variable error. *Chapter One* considers the importance of total survey error for social survey methodology and health services research. It examines in detail three components of nonsampling bias from

Kish's classification—field bias (respondent reporting error), processing bias (imputation error), and nonresponse bias and also provides estimates of standard error. The final section describes how components of the model are to be measured.

Part Two evaluates the effects of several types of error on social survey estimates of health services utilization, expenditures, health insurance coverage, and diagnoses for which a doctor was consulted. *Chapter Two* deals with survey error associated with social survey estimates of hospital admissions per 100 persons per year, hospital days per admission, percent of the population seeing a physician within a year, and mean number of physician visits per year for those who saw a physician. Of particular interest is the extent to which differences in medical care use by demographic characteristics and perceived and evaluated health levels are affected by survey error. Field biases tend to be smaller than the standard errors of social survey estimates of hospital admissions and length of stay, yet they are generally larger for estimates of the percent of the population seeing a physician and for the mean number of physician visits.

Chapter Three focuses on the accuracy of social survey estimates of expenditures for hospital admissions, physician office visits, and visits to emergency rooms, outpatient departments, and clinics. Approximately 80 percent of all persons reporting expenditures for either hospital or physician services were inaccurate to some degree. In addition, our imputation procedures underestimated expenditures for emergency room, outpatient, and clinic visits. Both field and processing biases are much more likely to exceed standard errors of survey estimates for expenditures than for measures of use.

Chapter Four examines survey error in estimates of family premiums paid for health insurance, the services covered by a policy (including hospital, surgical-medical, doctor visit, major medical, prescribed drug, and dental), and various sources of payment for hospital admissions (such as out-of-pocket, voluntary insurance, Medicare, and welfare). Families tended to report paying higher premiums than indicated by their employers and insurance companies, and respondents had difficulty in correctly identifying major medical coverage. Some of the largest field biases were observed for out-of-pocket and welfare payments per hospital admission.

Chapter Five discusses error in respondent reporting of illness conditions resulting in a physician's visit, conditions requiring hospitalization, and surgical procedures—all of which were matched with data provided by physicians and hospitals. Certain demographic characteristics (race, income, and residence) were associated with accuracy of reporting both conditions and surgical procedures, while measures of impact of illness (number and severity of conditions, length of stay, and level of expenditures) were associated with accuracy of reporting conditions only; self- versus proxy reporting was not found to be related to accuracy of reporting.

Chapter Six designates certain categories of respondents and certain physicians contacted during verification as "problem respondents." Persons who refused permission to verify data were likely to be uncooperative in other areas as well (such as reporting income); physicians' cooperation was influenced most by the number of patients for whom they were asked to provide data. In relation to under- and overreporting, persons who reported one visit in the social survey were more likely to have underreported if they were inaccurate, while persons reporting five or more visits were more likely to have overreported. Self-respondents were found to be slightly more accurate than proxy respondents.

Part Three describes alternatives to the approaches used to assess bias in the previous part and in earlier reports (Andersen, 1975; Andersen, Kasper, and Frankel, 1976). *Chapter Seven* compares estimates based on three approaches to noninterview adjustment—a process that involves choosing categories, computing response rates within each, and deciding on values to impute to those who were not interviewed. The approaches include the one used in the 1970 CHAS survey and two alternatives, one based on geographic location and the other on using information about the reason no interview was obtained. All three sets of estimates are similar for population estimates, mean physician visits, and mean expenditures for hospital admissions.

Chapter Eight examines alternative approaches that may be followed when a respondent is unable or unwilling to provide answers to some questions. It is argued that imputing missing data on a case-specific or subgroup-specific basis is actually more conservative than excluding missing data cases or items from analyses. The effect of imputation on total hospital expenditures is discussed in detail, and use of external and internal data for purposes of imputation are compared.

Chapter Nine describes two types of weighting used with the 1970 CHAS survey, the first of which adjusts for unequal selection probabilities. The reliability of various estimates is compared with that expected from a self-weighting sample costing the same. It is not clear which procedure—the one resulting in the CHAS sample or the one leading to a self-weighting sample—is preferable, for the relative importance of various estimates must be considered. The second type of weighting is post-stratification, which adjusts a sample's distribution to the distribution found in more reliable data. Data resulting from two different post-stratification adjustments are compared with each other and with data unadjusted in this way. When considering estimates of physician visits and hospital expenditures, the three sets of estimates are quite similar. A more thorough investigation of the expected effect of post-stratification on various estimates is suggested.

Chapter Ten discusses several models for developing survey designs that would minimize total survey error. In all models, it is assumed that both social survey data and data from a record check procedure are available. Optimal allocation of the total budget between social survey and record check procedures and procedures by which the researcher makes use of the social survey and verification data are two of the issues considered in relation to each model.

In Part Four, *Chapter Eleven* reviews the major findings and suggests implications for survey research in general, health services research in particular, and social policy regarding health services delivery. More than $1 million were spent to collect and analyze the data used in this study. We estimate that verification used one third of these resources, extended the total data collection time by about eighteen months, and greatly complicated the data processing and analysis. Verifications in other large-scale social surveys may require a similar effort. Consequently, a crucial question considered in Chapter Eleven is whether verification seems justified in future studies. The answer depends partly on assessing how much estimates change after incorporating verification data. If the overall effort seems worthwhile, then cost-efficient ways of conducting a verification and effective methods for adjusting estimates using verification data become important. We cannot presume to answer such questions definitively, but we do hope the documentation of our experience will prove useful to others concerned with the validity of social survey estimates.

Acknowledgments

This report was funded by Contract No. HRA 106-74-24 from the National Center for Health Services Research (NCHSR) and co-sponsored by the National Center for Health Statistics (NCHS). Our first project officer, Daniel Walden of NCHSR, was not only instrumental in facilitating the contract but also served the study well in a collegial role by helping to formulate the questions to be addressed. We also wish to thank Robert Fuchsberg, who served as co-project officer for NCHS, and William Lohr, who took on the day-to-day responsibilities of project officer for NCHSR in a supportive and congenial fashion shortly after the contract was awarded. We also wish to thank the Center for Health Administration Studies and the National Opinion Research Center of the University of Chicago for "free" time and additional resources necessary to complete this project. Special thanks are due Kenneth Prewitt, director of NORC until July 1, 1979, whose support facilitated the publication of the text tables.

Joanna Lion was the original project director for the study before she became Director of Information Services at the Massachusetts Hospital Association. She played a major role in designing the study and writing initial project reports. We also appreciate her efforts as a manuscript reviewer, for they have added to whatever clarity this volume possesses.

Nancy Persechino provided much of the laborious research assistant work that is necessary in a project of this type. She was assisted by Alan Muenzer, Dale Jaffe, and Larry Corder. A great deal of custom programming was needed for some of the analyses included within. We feel fortunate to have been assisted by an able group of programmers including Robert Cripe, Frank Novak, and George Yates.

The bulk of the typing for the several drafts of this report was conscientiously done by Linda Randall and Joyce Van Grondelle. We thank other Center for Health Administration Studies members: Nancy Hague, Sharon Kulikowski, Margarita O'Connell, Charles Sellers, June Smith, June Veenstra, and Patricia Winning —all of whom willingly pitched in when necessary. The aid of Susan Campbell at NORC was invaluable in the preparation of the text, the accompanying tables, and the index.

Our co-authors, Martha J. Banks and Virginia S. Daughety, not only made contributions to this volume through their own chapters but also provided substantial assistance for other chapters as well. Both also competently assumed many of the coordinating duties prerequisite to seeing this volume in print.

Kent Marquis of RAND Corporation directed many helpful suggestions toward making our work easier to understand and more useful. We feel fortunate indeed that Seymour Sudman of the University of Illinois was willing to serve as the technical reviewer for this volume. His careful work has no doubt saved us much embarrassment; however, we bear sole responsibility for any blunders that might remain.

Chicago, Illinois
July 1979

RONALD ANDERSEN
JUDITH KASPER
MARTIN R. FRANKEL

Contents

Part Two: Estimates of Survey Error

Part Three: Alternative Approaches to Improve Survey Estimates

Part Four: Summary and Implications

Tables

✲ ✲ ✲ ✲ ✲ ✲ ✲ ✲ ✲ ✲ ✲ ✲ ✲ ✲ ✲ ✲ ✲

The Authors

✲ ✲ ✲ ✲ ✲ ✲ ✲ ✲ ✲ ✲ ✲ ✲ ✲ ✲ ✲ ✲ ✲

Ronald Andersen is associate director of the Center for Health Administration Studies (CHAS) and professor in the Graduate School of Business and the Department of Sociology at the University of Chicago. He was study director of the nationwide social survey and verification that produced the data for this book. His major research interests include social survey research methods, utilization of health services, and international comparisons of health service systems. Among his publications are *A Behavioral Model of Families' Use of Health Services* (1968), *Medical Care Use in Sweden and the United States* (1970), *Development of Indices of Access to Medical Care* (1975), *Equity in Health Services: Empirical Analyses in Social Policy* (1975), and *Two Decades of Health Services: Social Survey Trends in Use and Expenditure* (1976). He received his Ph.D. degree in sociology from Purdue University in 1968.

MARTHA J. BANKS is sampling project specialist at the Center for Health Administration Studies and a doctoral candidate in the Department of Sociology at the University of Chicago. She previously was associated with the National Opinion Research Center and was a mathematical statistician at the Bureau of the Census. She was senior author of a paper on standard errors of the Current Population Survey, which was presented to the American Statistical Association in 1971, and wrote a portion of the methodological appendix of *Two Decades of Health Services: Social Survey Trends in Use and Expenditure* (Andersen, Lion, and Anderson, 1976). She also presented a paper to the American Statistical Association in 1977, "An Indication of the Effect of Noninterview Adjustment and Post-Stratification on Estimates from a Sample Survey," based on Chapters Seven and Nine of this volume.

VIRGINIA S. DAUGHETY is policy analyst with the Center for Health Services and Policy Research, Northwestern University. At the time this book was written, she was research specialist at the Center for Health Administration Studies, University of Chicago, where she was instrumental in the development of physician norms to determine appropriate access to medical care for specific illnesses. She coauthored "Psychologically Related Illness and Health Services Utilization" (*Medical Care,* Supplement, May 1977). She received her M.S. degree in health planning from the University of California at Los Angeles in 1975.

MARTIN R. FRANKEL is an associate professor of statistics, Bernard Baruch College, City University of New York. His research interests are in the area of survey sampling. Frankel's publications in this field appear in the *Journal of the American Statistical Association, Journal of the Royal Statistical Society,* and the *Journal of Marketing Research.* He received his Ph.D. degree in mathematical sociology from the University of Michigan in 1971.

JUDITH ANN (DELLINGER) KASPER is a service fellow at the National Center for Health Services Research. She is presently an analyst on the research team for the 1977 National Medical Care Expenditure Survey (NMCES), which plans to address such health policy issues as costs of various national health insurance proposals, access to medical care, and costs of episodes of chronic and acute illnesses. As a research associate at the Center for Health Administration Studies, University of Chicago, she served as study director of the project that produced this volume. Her previous publications and dissertation focused on family size and children's use of physician services, based on the 1970 CHAS data. She received her Ph.D. degree in sociology from the University of Chicago in 1976.

Total Survey Error

✲ ✲ ✲ ✲ ✲ ✲ ✲ ✲ ✲ ✲ ✲ ✲ ✲ ✲ ✲ ✲ ✲

Applications to Improve Health Surveys

CHAPTER 1

A Model of Survey Error

Ronald Andersen
Judith Kasper
Martin R. Frankel

In 1944 Deming listed thirteen factors that affect survey usefulness. Among these were response variability, bias from nonresponse and late reports, sampling errors and bias, and such elements as errors in interpretation of data and imperfect questionnaire design. A few years later Hansen and others defined "error" in a survey result as "the difference between a survey estimate and the value which is estimated" and "response error" as a designation for nonsampling errors "introduced during the course of data collection" (1951, p. 147). Processing errors attributed to coding, editing, and calculating and nonresponse errors also were defined in these early publications. These definitions seem to have gained widespread acceptance among survey researchers, and most studies of error since have focused on elaborating sources of response error.

The 1945 Census of Agriculture contained the first measurement of response errors incorporated in a regular census (Mauldin and Marks, 1950). Since then the Census Bureau has continued to modify survey methodology, attempting to reduce nonresponse, coverage, and response errors through reinterviews, comparison with outside data sources, and other strategies. (See especially the series of technical papers and working papers produced by the Census Bureau, for example, Working Paper No. 36, *Response Variance in the Current Population Survey.*) In addition to these efforts to reduce error in the data-collecting process, there are many studies on the interviewer's characteristics and interaction with the respondent as sources of response error (Allen, 1968; Cannell and Kahn, 1953; Marquis, 1969; National Center for Health Statistics, 1972a), problems in questionnaire design (Belson, 1968; Marquis, Marshall, and Oskamp, 1972; Nisselson and Woolsey, 1959; Suchman and others, 1958), and difficulties with recall and interview content (Cartwright, 1963; Dakin and Tennant, 1968; Gray, 1955; Knudsen, Pope, and Irish, 1967; Larson, 1958). However, there have been few attempts to evaluate the effects of these errors on estimates from a large-scale social survey, as some researchers have advocated (Ferber, 1966; Hansen and Waksberg, 1970).

Interest in nonsampling errors in health services research is more recent. Nonsampling errors and attempts to measure them have been the subjects of several publications from the National Center for Health Statistics (NCHS). These attempts to assess error include changing interviewers and reinterviewing respondents in studies of hospitalization costs and experience and checking records against interview responses (National Center for Health Statistics, 1965b, 1965c, 1973a). At a recent national conference jointly sponsored by NCHS and the National Center for Health Services Research (NCHSR), the issue of total survey error was a major organizing theme (National Center for Health Services Research, 1977). Participants discussed ways to increase "the validity and reliability of measures of health" and it was pointed out that "questions of validity and reliability often focus upon sampling errors as the major measurement issue, whereas nonsampling measurement errors are equally important" (1977, p. 37).

Total survey error thus has been identified as an important problem in the literature of both survey methodology and health services research. To study the problem empirically, some theoretical model is necessary. The model used in this volume is described in the next section.

A Model of Total Survey Error

To assess the magnitude of total survey error in survey estimates, various sources of error and ways to measure them must be defined. A rebirth of interest in evaluating total survey error has led to several attempts to develop models of error sources that can be applied to surveys in a general fashion. Most of the models developed to evaluate total survey error are generalizations of the mean squared error formulation used in mathematical statistics. This formulation is too limited for our purposes because it assumes perfect implementation of the sample design, perfect measurement of the variable values for all sample elements, and perfect sampling frames. Thus it allows only for "statistical bias" and sampling variance. Under the simplest case this model assumes that a well-defined population of elements exists, each of which possesses a value for the variable of interest. The true population parameter is defined in terms of this variable, measured without error, over the entire population. Examples of often used population parameters are population means, sums, and proportions.

In the model of mean squared error, the survey estimator is evaluated in terms of its mean squared deviation from the true population parameter, over the set of possible samples that may be selected under a particular sample design. Using the sample mean as an example,* if we let $\overline{Y}_{\text{true}}$ denote the value of the true population mean and $\bar{y}_c$ denote the estimate of the population mean that would be developed from the cth possible sample, then the squared

*If all samples are not equally likely, this mean is weighted by the relative probability of sample selection. This weighting for differential probability, when required, is assumed throughout.

deviation or difference between the cth sample estimate and the population parameter being estimated is

$$(\bar{y}_c - \bar{Y}_{\text{true}})^2$$

The mean squared error of the sample estimator $\bar{y}$ is defined as the average or the mean of the squared deviation between the sample estimate and true population parameter over all possible samples, for a specific design. This is denoted as

$$\text{Mean Square Error } (\bar{y}) = E\ (\bar{y}_c - \bar{Y}_{\text{true}})^2 \qquad (1)$$

where the letter "E" in front of the expression for the squared deviation indicates we are taking the average or mean or "expectation" over all possible samples that could be selected under the specific sample design.

It can be shown mathematically that the single-term definition of the mean squared error given in (1) is equal to the following expression involving two terms:

$$MSE(\bar{y}) = E(\bar{y}_c - \bar{Y}_{\text{true}})^2 = E(\bar{y}_c - E(\bar{y})\ )^2 + (E(\bar{y}) - \bar{Y}_{\text{true}})^2 \quad (2)$$

Both terms make use of the symbol $E(\bar{y})$, which is simply the expected or mean value of the sample estimate $\bar{y}$ taken over the set of samples possible under the sample design. The first term $E(\bar{y}_c - E(\bar{y})\ )^2$ is referred to as the sample variance and is the average over all possible samples of the squared deviation between a specific sample mean and the expected sample mean. The second term is the squared difference between the expected sample mean and the true population mean. This latter term $(E(\bar{y}) - \bar{Y}_{\text{true}})^2$ is called the squared bias of the estimator $\bar{y}$ with respect to parameter $\bar{Y}_{\text{true}}$.

Although this formulation of mean squared error is too limited for our use, it does convey the basic principle that will be central to our discussion of total survey error. In making use of expression (2), we have partitioned the mean squared error into two components. One of these components, the sampling variance, will be generalized to include all error sources that result in a difference between an estimate based on a particular or specific sample and the results that would be obtained if we were to average over all possible samples. The second component, the squared statistical bias of the survey estimate, will be generalized to include all error sources that would produce a difference between the average, taken over all possible samples, of the survey estimator and that of the true population parameter for which the estimate is sought.

Leslie Kish (1965, pp. 509–527) described the general model used in our discussion. Kish's model is an attempt to synthesize a number of other formulations (Cochran, 1963; Hansen, Hurwitz, and Madow, 1953; Sukhatme, 1954; Zarkovich, 1963) into a model possessing both conceptual simplicity and generality. Kish uses the concept of "individual true values of a variable for each population element" as a point of departure (1965, p. 514). This concept, which Kish notes was developed by Hansen, Hurwitz, and Madow (1953), must satisfy three criteria: (1) the true value must be uniquely defined; (2) the true value must be defined in such a manner that the purposes of the survey are met; and (3) where it is possible to do so consistently with the first two criteria, the true value should be defined in terms of operations that can actually be carried through (even though it might be difficult or expensive to perform the operations).

Specifying the Components in Kish's Survey Error Model

As a first step in applying this model to actual surveys, we use the perspective supplied by Kish and illustrated in Figure 1, which applies a classification scheme to the various sources of bias and variable error. (The actual formulas used in calculating biases are given in Appendix A and discussed in the relevant chapters.) Bias, according to Kish, "refers to systematic errors that affect any sample taken under a specified survey design with the same constant error" (1965, p. 509). Variable errors are assumed to be random.

Bias

There may be two types of bias: sampling and nonsampling. Sampling biases include those resulting either from the sampling process itself or from the statistical estimation process. Of all sam-

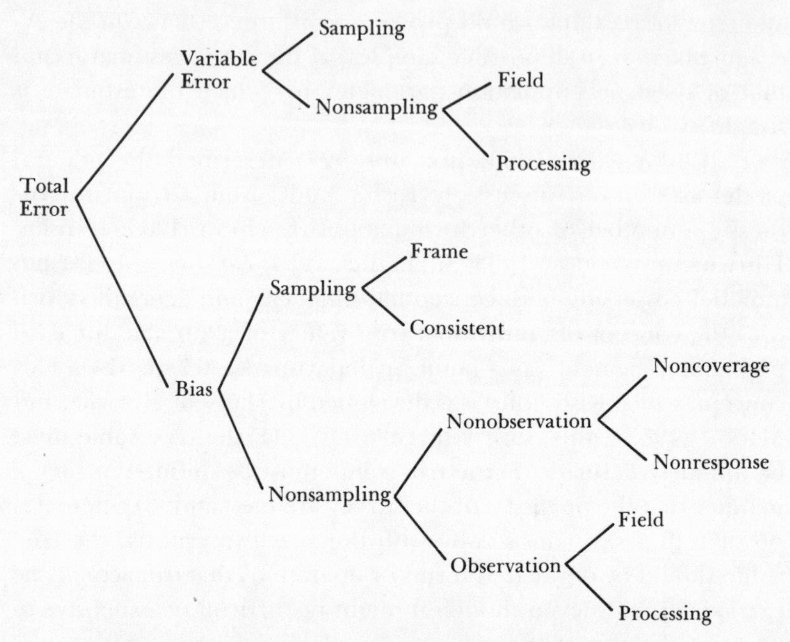

Figure 1. Classification of Sources of Survey Error

pling biases, those attributable to sampling frame sources are probably the most serious. Frame biases are the consequence of using inappropriate selection procedures to choose a sample from the universe. For example, frame bias would occur if each element in a list were given an equal chance of selection yet some population elements appeared more than once on the list. Consistent biases, caused by biased but consistent estimators, decrease with increasing sample size and vanish from the population value in a 100 percent sample (Kish, 1965, p. 519). For example, there is a consistent sampling bias associated with using the variance of a sample to estimate the variance of a population or using a ratio from a sample to estimate a population ratio.

Nonsampling biases probably account for the largest source of total survey error. Both the survey analyst and the survey user often fail to recognize their existence and impact. Kish further distinguishes between nonsampling biases of observation and nonobservation. "Biases of observation are caused by obtaining and recording observations incorrectly" (1965, p. 520). These again may be divided into two classes: field (response) biases, which arise in the collection of observations and include interviewing, enumerating, counting or measuring problems, and processing biases, which occur while translating the data into machine readable form and analyzing it.

Nonobservation biases that arise from failure to obtain observations on some segment of the population are also of two types, nonresponse and noncoverage. Nonresponse refers to the failure to obtain observations on some elements selected and designated for the sample. This failure may be due to refusals, respondents who were not at home during the survey period, or lost questionnaires. In contrast, noncoverage denotes failure to include some elements of the defined survey population in the actual operational sampling frame. This bias due to noncoverage may appear similar to the "frame" bias; but it differs in that the element's probability of selection is zero rather than some nonzero, but incorrect, value.

Variable Errors

Variable errors also are of two types: sampling and nonsampling. Variable sampling errors arise because only a subset of the population can be in the sample. These sampling errors differ from sampling biases in that they have zero expectation; in other words, averaged over all possible samples, their net effect vanishes. In the CHAS-NORC survey the multistage stratification and weighting procedures are susceptible to variable sampling error.

Variable nonsampling errors are those sample-specific errors that arise from sources other than sample selection and statistical estimation. Nonsampling variable errors may stem from the respondent, the interviewing process, supervising, coding, editing, and data entry. A low reliability coefficient produced in a test-retest of a survey questionnaire on the same population indicates variable nonsampling error.

Measuring Components of the Model

The first step in applying Kish's model to our survey (which is described in the Preface) is to make its components operational.

We will not attempt to measure each component shown in Figure 1 but will focus on three components of nonsampling bias—field, processing, and nonresponse. We also provide estimates of standard error, a type of variable error. (For further information about standard errors for CHAS-NORC survey variables, see the methodology section of Andersen, Lion, and Anderson, 1976).

We have chosen to measure sources of bias and error that may significantly influence the validity and reliability of data and that our survey is particularly suited to evaluate. It is not possible to measure each error component in its entirety, so their sum may not reflect the best available estimate of the "overall" nonsampling bias. These possible discrepancies between the sum of the "main effects" of the three bias components and the overall nonsampling bias would be due in part to the interaction among the "unmeasured" portion of each of these three components and in part to the nonmeasurement of the fourth nonsampling bias component—noncoverage. We attempted to eliminate a portion of the differential noncoverage as well as nonresponse biases by post-stratification (ratio estimation) adjustment on the basis of sixteen strata defined by family size, family income, race, and location inside or outside a Standard Metropolitan Statistical Area (SMSA). (See Chapter Nine for a more detailed account of the weighting procedures.)

In the following chapters we will be assessing the effect of these bias components on group means, which reflect the demographic and health status characteristics of the population, and on the total sample mean. Negative biases in the tables that follow indicate the social survey estimate is too high; positive biases indicate the estimate is too low. We adopted this convention so the reader can easily calculate estimated true values by adding or subtracting biases from the original estimates. We also will compare differences between group means before and after adjustment for biases. A description of our measures of variable error (standard error), nonresponse bias, field bias, and processing bias follows.

Standard Error

Our only estimates of variable error are those derived from applying standard error formulas, which reflect both the clustered and stratified nature of the sample design and were produced using a technique described by Nathan Keyfitz (1957). Although standard errors are usually represented as measuring the sampling portion of variable error, the standard error terms in fact reflect portions of the nonsampling variable error as well. These nonsampling errors typically include random errors in respondent reporting and a correlated response variance that reflects differences in the performances of individual coders, editors, interviewers, and other survey personnel. Because there is evidence that nonsampling variance may be sizable in some cases, the absence of a direct measure of this error source in our calculations of total error should be noted (Bailar, 1976; Bailey, 1975; Koch, 1973).

Nonresponse Bias

Nonresponse refers to the failure to obtain observations from persons in the designated survey sample. Persons may not have been at home, may have refused to cooperate with the survey, may have been incapable of responding, and so on. Estimating the impact on survey results had information from nonrespondents been obtained is a two-step procedure: First, the percentage responding and not responding must be calculated for the categories of the independent variables of interest; second, rates of hospitalization or physician use (the dependent variable) must be calculated for nonrespondents (the survey provides rates for respondents). The second step involves extrapolation from other data sources, specifically, studies that give information about nonrespondents' behavior. Admittedly, we have made some rather heroic assumptions in applying data from independent sources to make estimates of nonresponse bias in this book (see Chapters Two and Three). The problems this method entails should be weighed against the advantages of being able to include information for these people in the data. Alternative methods of estimating nonresponse bias are presented in Chapters Seven and Eight. Although both demographics and measures of health level are used as independent variables in this work, we have calculated nonresponse biases only for demographic variables (Chapters Two and Three) because independent estimates of utilization of nonrespondents according to the health characteristics were not available.

Information from special screening and selection techniques used to increase the yield of persons sixty-six years of age and older and low-income families was used to estimate response rates by age of oldest family member and family income. (See Chapter Nine for a discussion of the overall sampling design and these special techniques.)

No information was collected on nonrespondents' race, but data correlated with race were collected on both respondents and nonrespondents. Using the respondents' data, we developed an estimating equation by regressing race on measures of residence, region, dwelling type, and income level. The regression weights from this equation were then used to estimate the proportion of nonwhites among nonrespondents, and response rates for whites and nonwhites were estimated.

Residential information was collected on both respondents and nonrespondents from census classifications and interviewers' observations. This information was then used to estimate response rates for residential categories. The estimated response rates for age of oldest family member, near poverty level, race, and residence derived from the described procedures are shown in Table 1.1.

The second step in calculating a nonresponse bias involves obtaining information about nonrespondents' behavior for the dependent variable of interest. Because this procedure varies with the dependent variable, it will be explained in each chapter where nonresponse biases are presented. It is important to calculate nonresponse bias because of the possibility that its effect on one category mean may be different from that on the other. As Kish points out, "Distinct from the usual vague guesses are some formal attempts to estimate the nonresponse bias, and to reduce it by incorporating that estimate into the statistics" (1965, p. 557).

Field Bias

Field bias arises from the difference between the information given to us by survey respondents and the data collected from verification sources. Information from doctors, hospitals, and insurance companies in this study is generally accepted as the "true" value (Hansen and Waksberg, 1970, p. 318) and discrepancies with survey information are attributed to respondent reporting error. We know there are errors in record sources as well (Mathews, Feather, and Crawford, 1972). For simplicity, however, our working assumption is that the verification sources of this study are generally more accurate than the respondents with respect to the key dependent variables.

Respondents make several types of error. The most common reporting error is a disagreement between the respondent and the verification source about the number of days, number of visits, costs, or insurance payments for the patient.

Another type of error was a respondent's report of use of services or expenditures that was rejected by the verification source (false positives). Frequently in these cases the care or expenditures the respondent reported did occur, but not during the survey year.

A third type of error is the failure to report information that was later discovered in verification. For example, some persons reported no hospitalizations but reported seeing a physician. Upon contacting the doctor, we found the physician had hospitalized the patient. These cases are frequently referred to as cases "discovered" in verification. We alluded earlier to the fact that although the verification process allowed discovery of some unreported information, it was not optimally designed to identify respondents who incorrectly reported no visits to physicians or no hospital admissions because verification was not attempted on these "false negatives." Theoretically, this design will lead to an inflated estimate of respondent overreporting of incidence of events, for example, number of hospital admissions or physician visits; but it should provide less distortion in estimating magnitude of events, such as average length of hospital stay or mean expenditures per hospital admission. Care the respondent did not mention was discovered only if a mentioned source reported it. For example, an insurance company the respondent identified might report a claim for a hospitalization the respondent had not reported or a hospital confirming a reported hospitalization might indicate a physician the respondent had not reported as having provided care had performed surgery.

Persons who refused to allow verification presented a special problem. Approximately 10 percent of the families in the social survey refused to sign the permission forms needed to obtain information from doctors, hospitals, and insurers. Our solution was to adjust the survey information these individuals provided by the amount of error for cases where we received a report from both social survey and verification sources. If persons who refused to allow verification were trying to hide information or were generally less accurate than others, then ours is probably a conservative estimation of the error for this group. We did not include rejected-in-verification or discovered-in-verification cases in our adjustment factor because our verification procedure made rejection much more likely than discovery of care for expenditures. (See Appendix A for treatment of no verification cases in bias calculations.)

Table 1.2 provides some idea of the effect of the verification process on the number of hospital admissions, physicians seeing survey respondents, and health insurance policies. Column 1 shows the total number of each event reported by families. Column 2 indicates the number of events for which verification was not attempted either because permission to verify was refused or identifying information was inadequate. The number of cases for which verification was attempted but not completed due to nonresponse or refusal to cooperate is given in column 3. Events the family reported for which verification was completed are shown in columns 4 and 5. If the verification source denied that the service the family reported had been provided or indicated the event occurred outside the calendar year, the observation was rejected (column 4). If the verification source confirmed the family's report, the observation was accepted (column 5). In a few instances the family report was accepted and is found in column 5 even though the verification source did not confirm the event. This happened, for example, when the respondent reported services for which he or she had documented records but the provider appeared to have inadequate records or had made an incomplete records check.

In summary, the field bias consists of four subcomponents: (1) an event such as a hospital admission reported in both the social survey and verification but with days or expenditures about which there may be disagreement, (2) rejected social survey reports, (3) events discovered in verification, and (4) events reported in the social survey for which no verification was obtained. The bias observed when verification was obtained was attributed to these events. These subcomponents combine to form the overall field bias and are generally not reported separately. However, they will be reported when we judge them to be of substantial help to the reader in understanding the reporting bias.

Processing Bias

Although Kish identifies several sources of processing bias, our operational definition of the term is imputation bias. Whenever survey respondents reported physician contact or a hospital stay but were unable to report the number of days, the length of stay, or the expenditures involved, we imputed this information. Therefore a potential source of bias was introduced during the office processing phase of the survey in the effort to make the data more complete.

Outside data sources were the basis for imputation. For example, for physician visits and related expenditures, physician data provided by the American Medical Association according to specialty, geographic region, and urban or rural sites were prime sources used to impute missing respondent information. Similarly, a list of hospitals showing per-diem and outpatient charges from the American Hospital Association was used to impute hospital expenditures if the respondent failed to provide these data. (Chapter Eight contains a more extensive discussion of our imputation procedures as well as alternative strategies of item estimation or imputation.)

Imputation bias was calculated by comparing our estimates with the information verification sources provided later. Imputed data for persons who refused to allow verification were adjusted according to cases where information was imputed and verification was received (the same procedure used for these persons in calculating field bias).

Summary

Part One has described the general problem of survey error and the framework we propose to use for studying the problem. In Part Two we turn to specific analyses of survey error for various kinds of estimates made in health surveys. (The reader should consult the Preface for a description of the survey that produced the data analyzed in this volume and for discussions of both the organization of the book and the purpose of each chapter.)

In earlier publications using this data set, estimates based on the merger of social survey and verification information are referred to as "best estimates" (Andersen, 1975; Andersen, Kasper, and Frankel, 1976; Andersen, Lion, and Anderson, 1976). These best estimates are generally less complete adjustments for survey error than the adjustments made in Part Two of this volume. The best estimates essentially substituted information from the verification when it was available for that from the social survey; they are not adjusted for nonresponse bias, imputation bias, or reporting bias (for those cases where no verification information was received).

CHAPTER 2

Health Services Use

Ronald Andersen

This chapter examines the survey error associated with four common measures of the population's use of health services: hospital admissions per one hundred persons per year, hospital days per admission, percentage of the population seeing a physician within a year, and mean number of physician visits per year for those who saw a physician. It suggests the extent to which differences in the use of health services might be affected by measurement error according to age, income, residence, and race. The relationship between respondents' perceived and evaluated health levels and the accuracy of our imputations of their utilization is also considered.

Other studies suggest that the nature of a person's illness (severity, number of diagnoses, perception of health) can influence whether or not a condition is reported and the details given about it (National Center for Health Statistics, 1965a, 1965c). For example, do persons in poor health, who worry about their health, or who have severe medically diagnosed conditions have more error associated with survey estimates of their utilization than do persons in better health? Finally, we will study differences in survey error according to the respondents' volume of health services use. Persons who use services infrequently may have special difficulty recalling instances when they did so, but repeated hospitalizations or medical appointments over an extended period of time may place extra burdens on persons attempting to re-create their medical care experiences. Because high utilizers account for a disproportionate share of all services consumed (those with five or more physician visits in a year are 24.9 percent of the population but account for 78.0 percent of all physician visits), large errors in their utilization estimates could seriously distort the picture of medical care consumption in this country.

Measurement of Variable Error and Bias

In addition to the standard error measurement, the three specific nonsampling biases in utilization estimates we will attempt to assess include (1) nonresponse bias resulting from differences between families interviewed and other families in the sample who refused to participate or were not interviewed for some other reason, (2) a field bias resulting from families who were interviewed giving incorrect or incomplete information about their utilization experience, and (3) a processing bias caused by differences between imputation of service units received by families who gave no indication of quantities of services used and the actual service units received by those families.

Nonresponse Bias

The response rates for various population subgroups were derived from a number of different sources as described in Chapter One. In addition to the response rate for each group (see Table 1.1), the utilization rates for both respondents and nonrespondents must be calculated in order to determine the nonresponse bias. Although utilization rates for response are available from study data, estimates of these rates for the nonrespondents must be calculated from other sources for each service being examined: hospital admissions, hospital days, percentage seeing a physician, and mean number of physician visits. These utilization estimates for nonrespondents are calculated for all subgroups defined by the demographic variables. No attempt was made to estimate the utilization rates of nonrespondents according to health correlates or level of use because of a lack of suitable independent data sources.

Hospital Admissions. The estimated admission rate for the noninterviewed population was based on a probability selection of 1,589 persons with one or more discharges from twenty-one short-term hospitals as determined from 1958–1959 hospital records (National Center for Health Statistics, 1965a). Respondents or proxies for them were interviewed about hospitalizations for the twelve-month period prior to the interview. Five percent of the patients were nonrespondents for the household survey and they were assumed to represent the universe of hospitalizations not reported in health interview surveys (National Center for Health Statistics, 1965a, p. 59). The universe of all nonrespondents (both those hospitalized and those not) for the same time period was estimated from the National Health Survey study of hospital discharges from short-stay hospitals for the United States from 1958 to 1960 (National Center for Health Statistics, 1962). The rate for nonrespondents was then adjusted upward for increases in hospital utilization between 1958 and 1970. For subgroup calculations, the ratio of the utilization of nonrespondents to the utilization of respondents was assumed to be the same as that for the total population. Although the twenty-one hospitals (and thus their patients) are not a probability sample of the hospitalized population, it was the best external data found for estimating nonrespondents' hospital experiences.

Average Length of Stay. The average length of stay for nonrespondents was estimated using data from the study of twenty-one short-term hospitals (National Center for Health Statistics, 1965a, p. 59). The mean length of stay for nonrespondents in that study was 8.8 days, compared to 7.4 days for respondents. The estimate for nonrespondents for 1970 was then adjusted to take into account changes in length of stay between 1958–1959 and 1970. For each subgroup the ratio of the utilization rate of the nonrespondents to the respondents was assumed to be the same as that for the total population.

Percentage of Population Seeing a Physician. Some evidence concerning the physician utilization experience of families who do not cooperate with health interview surveys is available from a study done of a seventeen thousand resident housing estate just outside London (Cartwright, 1959). Eighty-six percent of the people on the estate were registered with six general practitioners who recorded details of each of their consultations for one calendar year. Interviews were attempted with a sample of three sixteenths of the families concerning their health and medical care utilization. The nonresponse rate was 14.5 percent. The proportion of nonrespondents who saw a physician according to the general practitioners' records, adjusted by the differences between the proportion of persons seeing a physician on the estate and the proportion seeing a physician in the United States in 1970, was used as the estimate for U.S. nonrespondents.

No studies that examined the use of physicians by persons who refused to cooperate with a health interview were found for the subgroups considered in this study. However, physician utilization experience for persons who refused to participate in a health examination survey according to age, income, race, and residence is available (National Center for Health Statistics, 1969). In 1960–1962 a nationwide probability sample of the noninstitutionalized population ages eighteen to seventy-nine was drawn. Of the total sample of 7,710 persons, 86.5 percent were examined and tested. Prior interviews showed that the unexamined persons were less likely to have seen a physician in the last year than those examined. The experiences of the examined and unexamined corresponding most closely to the subpopulation groups considered in this chapter were used to adjust the estimate of percentage seeing a physician among all nonrespondents for each subgroup (National Center for Health Statistics, 1969, pp. 34, 44). It is tenuous to generalize from the Cartwright findings to the U.S. population or to assume that those who refuse to be interviewed are like those who refuse a physical examination. Still, it may be better to use these tenuous assumptions and make an adjustment for nonresponse than to assume implicitly there is no difference in behavior between respondents and nonrespondents.

Mean Number of Physician Visits. Although the London study showed nonrespondents were less likely to have seen a physician than were respondents, the mean number of visits for those seeing the physician was slightly greater for the nonrespondents than for the respondents (Cartwright, 1959, p. 359). The ratio of respondents' visits to nonrespondents' visits was assumed to be the same in the U.S. and London studies. Because the U.S. response rate is known, the mean number of visits for nonrespondents could be calculated. The estimated mean number of visits for nonrespondents in each subgroup was adjusted for the difference in proportion seeing a physician between that subgroup and the total population.

Field Bias

The field bias to be measured is operationally defined as the difference between the utilization of sample persons recorded in the social survey and that recorded in the verification studies of hospitals and physicians who provided care. There are biases included in both sources, but the verification data will be accepted as the "true" population value in this analysis.

The problem of underreporting generally increases with length of recall period, although respondents sometimes report events that took place prior to the bounded recall period (Cannell and Fowler, 1963; National Center for Health Statistics, 1965a). The respondent was asked to report on medical care received over a calendar year, so recall might be a particular problem in this study. However, certain study procedures were expected to reduce the bias. These procedures included asking separate utilization questions for each episode of illness and each provider within an episode. The importance of checking records before reporting medical experience was emphasized and crucial questions concerning hospitalizations and insurance coverage were followed up with the patient directly if the family respondent was unable to provide detailed information. Effects of these procedures are examined in several NCHS publications (1965b, 1965c, 1972a).

Some indication of the success of these procedures in reducing the overall recall bias can be noted by comparing the estimates from this study with those from studies that used shorter recall periods or records rather than respondent sources. For calendar year 1970 the mean number of physician visits per person (excluding telephone calls) as estimated by the NCHS using a two-week recall period was 4.1 visits compared to the CHAS social survey estimate of 4.0 visits. The NCHS estimate was reduced by 10 percent (the proportion of all NCHS provider contacts made by phone in that time period) to adjust for telephone calls, which are not included in the CHAS estimate (National Center for Health Statistics, 1972b, p. 3). The process of telescoping, that is, reporting visits that took place outside the calendar year, would tend to increase the CHAS estimate. NCHS estimates from a sample of short-stay U.S. hospitals show a discharge rate of 14.6 admissions and 117 days of care per one hundred persons per year for 1970 (National Center for Health Statistics, 1973c, p. 20). The comparable estimates for the CHAS social survey are 13.7 admissions and 128 days—a slightly lower admission rate but a higher number of days per one hundred persons. These comparisons suggest that recall bias has been considerably reduced in the CHAS study. (The formulas used to calculate the field or reporting biases for the social survey estimates are found in Appendix A.)

Processing Bias

When respondents did not know the length of a hospital stay or the number of physician visits, the office staff imputed values for these variables using rules based on such characteristics as diagnoses, type of facility, cost, and patient age. Later, verification data were obtained from hospitals and physicians on services the patient received. Differences between the office imputations and provider reports allowed us to calculate a processing bias. The calculation of this processing bias indicates whether the imputation rules used over- or underestimated actual utilization and whether any discovered bias would affect some subgroups more than others. (The general formula used for calculating the processing bias for utilization is found in Appendix A.)

Hospital Utilization Findings

The total number of admissions in the social survey was 1,771. Of these admissions, 163 were judged not to be valid admissions (mainly because they were found not to have occurred during the survey year); however 110 others were discovered as a result of the verification process. The net result is a negative bias; in other words, the adjusted admission rate is lower than that reported by respondents. It should be remembered that there were limited opportunities to discover stays not reported by respondents. The verification was obtained mainly through hospitals where respondents reported they had been admitted. Those hospitals sometimes reported additional stays the respondent had failed to report and occasionally indicated a stay for another member of the patient's family that had not been mentioned in the interview. In addition, some new stays were obtained from physicians who had been reported as providing nonhospital care to a sample member or through insurance companies that were mentioned for other reasons. However, for many sample members with no reported hospitalizations, such cross-checks were not available. Consequently, we must assume some bias in the discovery of new hospitalizations in the verification.

Table 2.1 shows that for two fifths of the admissions for which we received some information from the hospital, respondents and hospitals reported the same length of stay. For slightly over two fifths of the admissions the respondents reported longer lengths of stay than did the hospital. In one half of these cases the difference was one or two days; but for one quarter the social survey report was at least six days longer than that ascertained in the verification. In slightly less than one fifth of the admissions the length of stay the hospital reported exceeded that which the respondent reported. In over one half of these cases the difference was three days or more. In general, then, this table suggests that household respondents are more likely to overreport than underreport their average length of stay (assuming that the hospital records are correct), although a large minority are in total agreement with the verification source.

The lower half of Table 2.1 shows the percentage differences in length of stay between the social survey and the verification. It emphasizes that the social survey estimate is not only more likely to be higher than the verification estimate but also that when it is higher, the percentage difference is also likely to be greater. Thus, in almost one third of the cases where the social survey exceeds the best estimate, the difference is greater than 25 percent. The difference is practically never greater than 25 percent when the verification exceeds the social survey.

Table 2.2 shows estimates of hospital admission rates among various subgroups in the population and the effect of variable error and bias on these apparent differences. The variable error (standard error) is generally larger than the sum of the biases measured (nonresponse and field). The nonresponse bias appears consistent, negative, and relatively negligible for all population groups. It suggests that nonrespondents have fewer admissions than respondents and thus would reduce the total population rates if taken into account. (These inferences and those below concerning length of hospital stay should be tempered by remembering the rather heroic assumptions necessary to estimate the utilization rates for nonrespondents.) The field bias is also negative but is not consistent among subgroups in the population. People in older families, the poor, nonwhites, and the rural farm population were more likely to report admissions later rejected in the verification. Those who thought themselves to be in poor health or who worried about their health also had more admissions denied by the verification source. Similarly, people with multiple diagnoses and diagnoses physicians judged to be more severe had higher negative field biases than those with single diagnoses or diagnoses for which physician services were considered optional. Persons with poor perceived and evaluated health status had much higher initial hospital admission rates, which seems to be the major reason for their larger field biases. That is, the size of the field bias relative to the size of the initial estimate is not consistently larger for those in categories suggesting poor health.

The biases for estimates of length of hospital stay shown in Table 2.3 tend to be less than the standard errors, as was the case for admission rates. The nonresponse bias suggests that nonrespondents have longer stays when admitted, even though they may have lower admission rates, based on the results of our assumptions about nonrespondents' utilization rates. The field biases are mostly negative, indicating that people tend to overreport their length of stay in the hospital. An exception was the rural farm population, which slightly underreported. Few consistent differences in the field bias appear according to perceived or evaluated health status. It is difficult to make the case that severity of illness is systematically related to reporting errors concerning length of stay. Further, the level of reporting does not appear to be influenced by the number of days people spend in the hospital. Although those with the most days have the largest bias, the relative error for these admissions is no greater than for admissions of shorter duration. Processing biases are negligible in Table 2.3 because there were few cases where it was necessary to impute length of stay.

The main purpose of this analysis is to examine the effect of biases on the apparent differences among major subgroups in their utilization of hospital services. Table 2.4 shows the differences between the use rates of the first subgroup defined by a category of a given variable and all other subgroups defined by that variable. For example, the first unadjusted difference reported in Table 2.4 is admissions for persons in families with no member over sixty-five (13.3 per one hundred persons per year) minus the rate for persons in families including members sixty-five and over (20.7), which equals −7.4. The adjusted difference follows the same procedure but shows means corrected for nonresponse, field, and processing biases. The calculation for the above example is [13.3 + (−.1) + (−.3)] − [20.7 + (−.1) + (−1.5)] = −6.2. Thus, the adjusted difference between age groups is less than the difference between groups not considering biases. The standard error (SE) of the difference tests the statistical significance of the unadjusted and adjusted difference. It is calculated by the formula:

$$SE_{\text{diff}} = \sqrt{\left(\begin{matrix} SE \text{ of} \\ \text{mean}_x \end{matrix}\right)^2 + \left(\begin{matrix} SE \text{ of} \\ \text{mean}_y \end{matrix}\right)^2}$$

For the preceding example:

$$SE_{\text{diff}} = \sqrt{(.7)^2 + (1.6)^2} = 1.8$$

The unadjusted differences in Table 2.4 show that families with older members have considerably more hospital admissions and longer lengths of stay. Adjustment for the biases reduced slightly the difference in admission rates but had no effect on the differences in length of stay. The significance levels were not affected in either case.

Members of low-income families had somewhat higher admission rates and longer stays than members of higher-income families, according to the unadjusted estimates. The adjustment process reduced the admission difference and increased the length of stay difference. The significance level was not materially affected.

According to the social survey, whites had higher admission rates but there was no significant racial difference in length of stay. The adjustment for biases did not substantially alter the picture.

Comparison of the hospital experiences of individuals living in SMSAs outside the central cities (in other words, the suburbanites) with those living elsewhere showed few significant differences. The suburbanites seemed to have more admissions than central city residents and longer stays than urban residents not living in SMSAs. Adjustment for biases did not change these conclusions.

As expected, persons in better health according to both self-perceived and medically evaluated norms used fewer hospital services than those worse off, according to the social survey. We did not suppose that this general conclusion would be altered by the adjustment process, but we did suspect that the magnitudes of the difference between those with more and less severe health problems might change in some systematic way. There are minor changes in the magnitudes of the adjusted differences, but they do not appear systematic. Thus, we are unable to conclude that the estimates of hospital use for those with fewer health problems are more or less accurate than the estimates for those in poorer health. Similarly, differences in days per admission between those with short stays and those with longer ones are not altered by the adjustment process.

Physician Utilization Findings

Of the total sample, 67 percent reported seeing a physician during 1970. We were able to obtain verifying information from physicians on 63 percent of this group. As a result of verification, 7 percent of the persons who reported seeing a doctor were judged not to have seen a physician because the provider claimed no visit had taken place during calendar year 1970. In addition, .5 percent of the persons who said they had not seen a doctor were discovered to have had a visit because a physician who had seen another family member reported a visit for this person as well or because a claim for expenses for physician services was discovered for this person during our health insurance verification. That we were much more likely to deny care reported by the respondent than to detect unreported care must be considered a bias in our verification procedure, but we feel an analysis of percentage seeing a physician is useful in examining the extent to which verification of reported visits might differ among population subgroups. Further, our examination of number of visits per person is limited to average number of visits for those seeing a physician, hence reducing the bias resulting from our inability to discover persons who saw the doctor but reported no visits.

Table 2.5 shows that when we did receive confirmation of visits from physicians, the number of visits given by physicians matched that reported in the social survey for a little over one third of the patients. When the number did not match, the social survey estimate was more likely to be higher than the verification. Thus, patients appear more likely to overreport than underreport their mean number of visits. However, this conclusion must be viewed with caution because it depends on the completeness of physician reporting. If physicians and clinics tended to miss some visits for a patient in searching their records, we would overestimate the extent of patient overreporting. We have no evaluation of the completeness of the physician forms.

The sums of the biases are generally greater than the standard errors for percentage of the population seeing a doctor (Table 2.6). The nonresponse bias indicates that, overall, nonrespondents are less likely than respondents to see a doctor. However, there may be certain exceptions, in particular the nonwhite population.

The field bias implies, within the limitations of our verification process, that more persons report seeing a physician than was verified through third-party sources. The field bias appears to be highest for the nonwhite population and also relatively high for the rural farm population. Persons who perceive themselves to be in excellent health and who worry little or not at all about their health have higher reporting biases than those in poorer health and those who worry a great deal.

Table 2.7 shows that the biases for the mean number of physician visits, like those for percentage seeing a physician, tend to be larger than the standard errors of these estimates. The overall nonresponse bias appears negligible, and there is little variance among subgroups.

The field biases suggest that the number of visits reported is generally higher than the number verified and the bias is largest among demographic groups for those in older families, the poor, nonwhites, and those living in nonSMSA urban areas. Those with poorer perceived and evaluated health and those with higher numbers of visits have the larger negative field biases for number of physician visits. However, the relative size of the bias compared to the total number of visits does not appear systematically greater than for groups in better health and with fewer numbers of visits. The processing bias is generally negligible except for persons in poor health.

Table 2.8 shows unadjusted and adjusted differences in use of physicians among various population subgroups. The adjustment for biases increases the difference in percentage seeing a doctor between those in young families and those in families with elderly members but decreases the difference in mean number of visits for those seeing a physician. Similarly, the adjustment increases the gap between the poor and nonpoor in percentage seeing a doctor but decreases the apparent greater volume of visits by the poor.

The adjustment does not affect the difference between the poor and nonpoor in percentage seeing a doctor but suggests that nonwhites do not have more visits than whites once they make contact with the system. The relationship in physician use patterns between suburban dwellers and other residents does not appear to change substantially as a result of bias adjustments.

Turning to the health measures in Table 2.8, we find that adjustment for biases generally increases differences in percentage seeing a physician between those in excellent health and those reporting less healthy states but reduces the differences in mean number of visits between the excellent group and the rest. Similar patterns are noted in other health and level of use variables. Adjustment for percentage seeing the doctor by how much a person worries tends to increase the difference between those who do not worry and the other categories (an exception is the "none" versus "hardly any" comparison). Conversely, adjustments in differences for mean number of visits between those who worry least, have the least severe and fewest number of conditions, and low levels of use and those with more health problems reduce the discrepancy. The results then suggest that amount of illness does not have as much impact on number of visits as the unadjusted estimates would suggest.

Summary

This analysis of bias in social survey estimates of hospital and physician use suggests few changes in the hospital use differences between subgroups that would have policy implications. However, some potentially important changes in physician use estimates were observed. These have mainly to do with differences according to illness level. Generally, the differences in percentage seeing the doctor between those in good health and those in poor health are increased by the adjustment process. In contrast, differences in mean number of visits for persons seeing the doctor according to health level are decreased when adjustments are made for the biases.

CHAPTER 3

Expenditures for Health Services

Judith Kasper
Ronald Andersen

Getting accurate information from survey respondents about health service expenditures is generally regarded as more difficult than acquiring information about use. Third-party payers who may be billed directly by the hospital or physician complicate expenditure reporting, as does some persons' reluctance to reveal personal financial information. This chapter focuses on the accuracy with which expenditures for hospital admissions and for physician office, emergency room, outpatient department (OPD), and clinic visits are reported. As in Chapter Two, we are interested in the nonresponse and processing biases in social survey estimates of service expenditures.

Related Studies

Most studies of accuracy in reporting health experiences focus on the respondent's ability to report service use and physician or hospital contact. Details of the event including diagnoses and costs are presumed to follow once the event has been recalled. There are particular problems in reporting expenditures, however, especially when a third party pays large portions of a bill. Such problems have led some researchers to conclude, "it is difficult, if not impossible, for family members to report the amounts paid through insurance benefits, government programs such as Medicare and Medicaid, welfare, philanthropy, and other third-party payers" (National Center for Health Statistics, 1974, p. 1). (Chapter Four looks more closely at expenditure reporting in relation to specific types of third-party payers.) In our survey, for example, imputation of expenditures was required for 41.5 percent of the (unweighted) hospital admissions reported in the social survey because the respondent was unable to provide this information. Both the error in respondent reporting (field bias) and the imputation error (processing bias) in our study are affected by many respondents' inabilities to report expenditures; guesses by persons who do not know may increase the field bias and answers of "don't know" would increase the processing bias if the expenditures being estimated are atypical and deviate substantially from the standard used to make our imputations.

Using CHAS 1970 survey data, Kulley (1974) found that longer hospital stays and larger bills led to better reporting of the hospitalization occurrence, although not necessarily of details of the stay. Cannell and others (National Center for Health Statistics, 1965a) reported in a study of discharges during one year from twenty-one short-stay hospitals that the greater the financial strain associated with a hospitalization, the more likely it was to be reported on the first interview. Respondents were also asked what items they thought were too personal or prying. The largest single criticism referred to questions on finances, either about income or hospital costs. These studies suggest that magnitude of expenditures influences reporting of hospitalizations, but they say little about the accuracy of reporting expenditures themselves. Respondents' criticism in Cannell's study that finance questions are "too personal" supports the notion that persons may be reluctant to provide information about expenditures. This attitude may increase field bias if respondents choose to give a wrong answer to conceal this information; if they choose not to answer at all, more imputation will be necessary and processing bias may increase.

Marquis, Marquis, and Newhouse (1976, p. 926) argue that "surveys do not accurately estimate out-of-pocket expenditures for out-patient services whose costs are covered in more than small amounts by third parties and services whose costs are covered by the public sector." The fact that in our survey some expenditure imputation was made for 62.1 percent of the (unweighted) persons reporting visits to emergency rooms, out-patient departments, and clinics seems to support this argument. Less expenditure imputation was necessary for reporting office visits to physicians (16.8 percent needed imputation).

Nonresponse Bias

In addition to calculating the response rates for characteristics of population groups (Table 1.1), expenditure rates for respondents and nonrespondents are necessary to calculate a nonresponse bias. The nonresponse biases for expenditures incorporate the service use rates for nonrespondents arrived at in Chapter Two and thus are subject to the same limitations as the utilization nonresponse biases.

Expenditures per Admission

Average length of stay for respondents and nonrespondents was calculated in Chapter Two. A per-diem cost based on respondent information (total social survey expenditures for admissions/total social survey hospital days) was multiplied by nonrespondent average length of stay to produce nonrespondent expenditures per admission corresponding to respondent expenditures per admission. The response rate for population groups was adjusted for admission rates of respondents and nonrespondents; and a nonresponse bias—the difference in expenditures per admission had nonrespondent information been included—was calculated. Thus, this bias calculation assumes per-diem costs are the same for respondents and nonrespondents and differences in expenditures per admission are a function of estimated differences in length of stay and admission rates of nonrespondents.

Expenditures for Office, Emergency Room, Outpatient Department, and Clinic Visits

Nonresponse biases for physician expenditures were calculated in the same way as those for hospital expenditures. A cost per visit was derived from the social survey data and multiplied by the mean number of visits per person (this number differed for the respondents and nonrespondents, the nonrespondents' visit rate having been calculated in Chapter Two) to arrive at nonrespondent expenditures per person. The response rate for population groups was adjusted for differences in physician contact by respondents and nonrespondents and a nonresponse bias was calculated indicating the difference in physician expenditures had nonrespondent data been included.

Findings

Expenditures per Hospital Admission

For one fifth of all admissions (excluding 137 stays for which no verification was obtained), there was less than $1 difference between the social survey (which includes both reported and imputed values) and verification estimates of expenditures (Table 3.1). Another fifth of all admissions had errors ranging from $26 to $100. For 24 percent of all admissions, however, a difference of more than $200 separated the social survey and verification; for half of these the difference was greater than $400. The number of admissions in which the social survey expenditure estimate exceeded the verification estimate about equals that in which the reverse occurred.

The bottom half of Table 3.1 shows that for over a third of

all admissions the social survey expenditure figure was within 5 percent of the verification estimate. In 28.3 percent the difference between social survey and verification was greater than 25 percent of the verification estimate. As pointed out earlier, slightly more admissions were rejected in verification after being reported in the social survey (6.9 percent) than were discovered (4.8 percent). Generally, this table shows a discrepancy between the social survey and verification estimates of expenditures for an admission for close to 80 percent of all admissions. Many of these differences cancel out because there are almost as many cases in which the social survey is less than the verification as cases in which the social survey is more. Table 3.2 allows us to examine to what extent these differences are due to respondent reporting errors or errors of imputation and the effect of these differences on the mean estimates of expenditures.

According to Table 3.2, the largest expenditures per admission are for persons in families with a member over sixty-five years old, poor persons, and those who live in the central city or other urban areas of an SMSA. Total expenditures appear to be similar for whites and nonwhites. Those who perceive their health to be poor, worry a great deal, and have severe and multiple diagnoses also have higher expenditures per admission. The nonresponse bias is positive and usually larger than either the field or processing biases, suggesting the mean estimates would be higher had nonrespondent data been included (because we estimated the average length of stay to be longer for nonrespondents than for respondents).

Although the largest reporting errors generally occur for groups with higher mean expenditures, there are some differences in under- and overreporting. Nonwhites, those in older families, and persons in SMSA other urban areas have the highest levels of underreporting among the demographic variables. The highest levels of overreporting are among persons who perceive their health to be excellent or do not worry and persons with nonsevere diagnoses. In addition, persons with high expenditure levels in the social survey overreported and those with lower levels underreported. In general, healthier people tended to overreport and those who were less healthy, nonwhite, and in older families tended to underreport. The most accurate reporting was by younger families, whites, the nonpoor, and those who worry a great deal about their health. Field bias had virtually no effect on the total mean estimate because, as Table 3.1 showed, overreporting and underreporting tended to cancel out.

Imputation of expenditures for no answers tended to be slightly high. The total mean estimate is inflated by $7 as a result of processing bias. Imputation was least accurate for nonwhites, those in SMSA other urban areas, persons who perceived their health to be poor or worried a great deal, and persons with more severe diagnoses or higher expenditure levels. The expenditures imputed for admissions reported by all these groups were too high. One final point is that the standard errors are generally larger than the other error terms, suggesting these biases probably will not change relationships among the population groups.

Expenditures for Persons with Physician Office Visits

Table 3.3 shows that for 18.9 percent of persons with office visits the social survey and verification estimates of expenditures for these visits was approximately equal. For another 52.8 percent the difference between these estimates was $2 to $25. The bottom half of this table shows, however, that although the absolute differences are small relative to hospital expenses, in 18.2 percent of the cases the social survey was at least 75 percent higher than the verification estimate. For one third of the persons with office visits the social survey differed from the verification by 25 to 75 percent. This indicates that because expenditures for physician office visits are less than hospital expenditures, the actual magnitude of the difference between the social survey and verification estimates is apt to be relatively small. However, the relative error for physician expenditures is considerably larger than for hospital expenditures (an error of $5 for a $10 physician expenditure, for example, is off by 50 percent).

As in Table 3.1, the cases in which the social survey exceeds verification (41.3 percent) about equal the cases in which the verification is larger (39.8 percent). The field and processing biases are calculated slightly differently for physician expenditures than for expenditures for hospital admissions (see Appendix A). Expenditures for persons who had *reported* social survey data only were considered subject to field bias; expenditures for persons who had any imputed expenditures (whether or not they also had reported data) were considered subject to processing bias. This approach may underestimate field bias because some reported expenditures will be considered under processing bias. Processing bias may be slightly overestimated for the same reason.

Table 3.4 shows generally small biases of all types. The non-response bias is negligible except for nonwhites and persons in urban nonSMSA areas where the inclusion of nonrespondent data would increase the mean estimate for these groups. Persons who perceive their health to be poor or who worry a great deal and persons with high expenditure levels are the most serious over-reporters of office visit expenditures. Persons with low expenditure levels tend to underreport. Most other groups appear to be reasonably accurate reporters with a slight tendency to overreport their physician visit expenditures.

The imputation of expenditures for persons who reported visits but were unable to report expenditures appears to have been slightly high for most groups. The average processing bias for all admissions was $2 too high. Persons who perceived their health as poor, worried a great deal, or had four or more diagnoses were affected the most (the mean estimate for those who perceived their health as poor was $25 too high due to processing bias).

Comparing Tables 3.3 and 3.4 shows that for the majority of persons differences do exist between social survey and verification expenditure estimates. However, these differences are small in absolute terms and their effect on the mean estimates are small because they tend to cancel each other out (numbers of overreporters and underreporters are approximately the same, 41.3 percent and 39.8 percent respectively in Table 3.3).

Expenditures for Persons with Emergency Room, Outpatient, or Clinic Visits

Fifteen percent of all sample persons had expenditures for emergency room, outpatient, or clinic visits according to the social survey, the verification, or both. About one fifth (23.1 percent) of these persons had no difference between social survey and verification expenditure estimates for these visits (Table 3.5). For another two fifths (39.7 percent) the difference between the social survey and verification was between $2 and $25. Although the absolute differences in expenditure estimates appeared relatively small, the percentage differences were considerably larger. In a third of these cases the social survey report of visits (and corresponding report or imputation of expenditures) was rejected by the verification source as not having occurred. Many cases also were discovered in verification. One reason for this is the greater likelihood that Blue Cross or private insurance will pay for an emergency room or OPD visit than for a physician office visit. Most of the cases discovered in verification came from claims data. Unlike expenditures for office visits or hospital admissions, overreporting is much more likely than underreporting (46.3 percent compared to 30.6 percent) for emergency room, outpatient, or clinic expenditures. This suggests that field and processing biases will have a stronger influence on the mean estimates for these types of expenditures.

Table 3.6 indicates that although nonresponse bias is relatively small, it is most important for nonwhites and persons in urban nonSMSAs (as was true for office visits, Table 3.4) where inclusion of nonrespondent data would raise the mean expenditure estimate. The overall field bias for persons with emergency room, outpatient, or clinic visits suggests overreporting (particularly for those who perceived their health as excellent and had more than one diagnosis or high levels of expenditures); but average expenditure imputation was consistently too low. The field bias in this instance has less impact on the total mean estimate than the processing bias, which caused the total mean estimate ($44) to be $17 less than the verification suggests it should be.

Many more reports of clinic, outpatient, or emergency room visits than office visits or hospital admissions were rejected or discovered. Clearly, respondents' apparent overreporting of expenditures would have been much more serious had many visits not been discovered. The large number of rejected visits is probably due to a combination of respondent misreporting and poor record keeping by many emergency room and outpatient departments. The latter explanation, of course, reduces the validity of our estimates of respondent misreporting. The majority of persons (62.1 percent) required imputation for these types of expenditures. Verification of these cases showed our imputations were generally low. Among persons who reported all their expenditure information, however, one third of the visits were rejected and verification of other cases usually showed social survey reports of expenditures were too high. These findings seem to suggest it is difficult for respondents to report accurately expenditures for emergency room, outpatient, and clinic visits. In addition, our imputation method in this instance appears to have contributed substantially to survey error. The imputation method consisted of using the outpatient visit charge given by the hospital in the American Hospital Association survey and multiplying by the number of visits reported. When no outpatient charge was available, the average cost of hospitals of that particular type (psychiatric, long-term general, short-term voluntary, and so forth) was obtained from American Hospital Association tapes for 1970. The processing bias frequently exceeds the standard error of the mean estimate. However, without some imputation procedure it would be virtually impossible to estimate expenditures for emergency room, outpatient, or clinic visits given many respondents' inability to report this information.

Effects on Differences Between Population Groups

Tables 3.7 and 3.8 examine differences between social survey estimates for population groups before and after bias adjustments. Among the demographic characteristics, adjusting for biases in reporting hospital expenditures changes few conclusions concerning differences between groups, although the adjusted differences in expenditures between younger and older families do appear to increase. For the perceived and evaluated health variables adjusting for bias does not dramatically affect the relationship among categories. The difference in expenditures between persons who perceive their health as excellent and those who perceive their health as fair becomes a difference of only one standard error instead of two. The difference in expenditures widens, however, between those who do not worry and those who worry somewhat, those with elective only and those with elective and mandatory diagnoses, and those with one versus those with three diagnoses. The latter finding suggests these differences are more important than the unadjusted social survey estimates would lead one to believe.

Table 3.8 shows the same type of comparison for expenditures for persons with office visits and those with emergency room, outpatient, and clinic visits. The only instance in which differences in expenditures for office visits change significance levels after adjusting for bias is in the residence variable. The difference in expenditures between SMSA other urban and SMSA central city becomes less significant after adjustments have been made for field, processing, and nonresponse biases; the difference between SMSA other urban and urban nonSMSA becomes larger. Also of note is the reduction in difference between persons with large expenditures and those with small expenditures.

The larger field and processing biases for emergency room, outpatient, and clinic visit expenditures leads one to expect adjusting for biases to have a greater effect on social survey estimates. This seems to be what happens in fact. The differences between members of young and older families, the poor and nonpoor, and whites and nonwhites increase. The adjusted differences suggest that nonwhites and the poor have relatively higher expenditures for these types of visits than the social survey estimates unadjusted for biases indicate. These persons are also more likely to be Medicaid patients and may be underreporting their expenditures. Adjusting for biases shows that persons in younger families actually have higher expenditures for emergency room, outpatient, and clinic visits than persons in families with a member over sixty-five years of age.

Adjusting for the processing bias seems responsible for the increased significance of differences between the unadjusted and adjusted social survey estimates. Adjusting for the biases also increases the difference in expenditures between SMSA other urban and rural nonfarm residents and between SMSA other urban and urban nonSMSA residents. Among the perceived and evaluated health variables, however, the differences in expenditures among groups remain very significant (greater than two standard errors) whether or not social survey estimates of expenditures are adjusted for bias (the only exception is between persons who do not worry at all and those who worry hardly any). There is a general tendency for the bias adjustment to increase the difference between the group in the best health and that in the worst health. Generally, we conclude the impact of the biases is greater for demographic characteristics than for health status characteristics.

Summary

Approximately 80 percent of all persons who reported expenditures for either hospital or physician services were inaccurate to some degree. The overall tendency was to overreport expenditures, although certain groups consistently underreported, especially persons with low levels of expenditures. Nonwhites underreported emergency room, outpatient, and clinic visits and overreported physician office visits, as did persons who perceived their health to be fair or poor. This may reflect these persons' dependency on emergency rooms and outpatient services as regular sources of care and confusion about the distinction between seeing a doctor in a clinic and an office setting.

Our imputation procedures underestimated expenditures for emergency room, outpatient, and clinic visits but slightly overestimated physician office and hospital admission expenditures. Field bias was also slightly larger for emergency room, outpatient and clinic expenditures than for physician office visit expenditures. Consequently, the combined effect of processing and field biases changed mean estimates for emergency room, outpatient, and clinic visit expenditures considerably more than mean estimates for office visit expenditures.

Analysis of hospital admission and physician visit expenditures indicated that on a case basis instances in which the verification exceeded the social survey usually about equaled cases in which the social survey exceeded the verification. Errors in under-

and overreporting (or high and low imputation) per person tended to cancel out and have little impact on mean expenditure estimates. This is not the case for expenditures for emergency room, outpatient, and clinic visits. Biases affect the differences between mean estimates for several groups. Although social survey data seem less reliable in this area, the information from the emergency rooms and outpatient departments may be less valid than the information that hospitals provide about admissions or physicians about office visits.

CHAPTER 4

Health Insurance Coverage and Payments

Ronald Andersen
Virginia S. Daughety

The knowledge people have about their health insurance coverage and third-party payments for the health services they receive is important for a number of reasons. If they do not understand their benefit structures, they may not take advantage of covered services or may mistakenly believe they are covered for services that are not part of their benefit packages. Although it is often assumed that individuals lack knowledge of the details of their health insurance coverage, studies of the problem are limited (Blue Cross Association, 1975; National Center for Health Statistics, 1966).

The public's reporting about health insurance is also important from a systems perspective for estimating (1) the proportion of people with various kinds of coverage, (2) the relative importance of various third-party payers and consumers' out-of-pocket payments for medical care, and (3) how different population subgroups finance their medical care. The Social Security Administration (1976) and the Health Insurance Institute (1976) publish such estimates based on secondary data sources such as health insurance company records, hospital responses to the annual American Hospital Association's questionnaire, and income data from the Internal Revenue Service. However, detailed demographic and socioeconomic characteristics of the insured and uninsured population are available only from social surveys (National Center for Health Statistics, 1972b).

This chapter is concerned specifically with social survey biases in estimates of (1) family premiums for health insurance; (2) the types of services covered by health insurance; and (3) payments for health services by voluntary health insurance, Medicare, and welfare (including Medicaid), as well as patients' out-of-pocket payments.

Family Premiums for Health Insurance

The unit of analysis in this section is a health insurance policy that was in effect for one or more family members at some time during calendar year 1970. Included are plans designed to cover the costs of medical and dental care such as are provided by Blue Cross-Blue Shield plans; private insurance companies such as Aetna, Travelers, Metropolitan, and Mutual of Omaha; and independents such as Kaiser Permanente and Health Insurance Plan of New York. Excluded are plans that pay medical care costs only in connection with accidents and plans that provide supplementary income when the policy holder is sick or disabled and cannot work. Also excluded from this analysis is Medicare and Medicaid coverage.

Respondents reported 3,635 (unweighted) health insurance policies were in effect during 1970. Verifying information was obtained from insurance companies, employers, hospitals, and physicians on 2,745 policies. Of these, 181 policies were rejected because the verifying source indicated they were not health insurance as we defined it, the policy was not in effect during the survey year, or there was simply no evidence that the reported health insurance existed. In addition, 65 policies were discovered in the verification process. Included here are policies the respondent did not mention but a hospital or doctor reported as paying for care the patient received and policies in addition to those originally reported by the respondent that health insurance companies and employers listed when they were contacted about the reported coverage.

For each health insurance policy reported in the social survey, the respondent was asked for the premium the family paid for coverage during the survey year. Similarly, the insurance companies, employers, or both were requested to report yearly family premiums for each health insurance policy. The premiums the family actually paid should be distinguished from the total cost of the policy because the employer paid part or all of the total premium for 92 percent of the work group policies. (No analysis of the total insurance premium is possible because families were not asked to estimate employers' contributions to the cost of the insurance.)

Table 4.1 compares the premium amounts reported by the family with those given by the insurer from all policies for which we received information from both sources. For over one sixth of these policies the family and the verification source gave essentially the same amount. When a discrepancy was found, the family's report was more likely to be high. Table 4.1 suggests that the social survey estimate exceeded the verification estimate for about half the policies but the verification exceeded the social survey for about one third.

Table 4.1 further suggests that, in absolute magnitudes, the discrepancies are likely to be greater when the social survey estimate exceeds the verification estimate than when the verification estimate is the larger of the two. For example, in 5.5 percent of the policies the social survey estimate exceeded the verification estimate by more than $200; the verification estimate exceeded the social survey estimate by more than $200 in about 3 percent of the policies. Also, for about 10 percent of the policies the social survey estimate exceeded the verification estimate by more than 75 percent and for 3 percent of the policies the verification estimate exceeded the social survey estimate by more than 75 percent.

Table 4.2 shows the estimated field bias for premiums according to various characteristics of the family and the insurance policy. Nonresponse bias was not calculated because no external information was available on nonrespondents' premium reporting to social surveys. Processing bias was not calculated because no imputation for family health insurance premiums was attempted.

The overall field bias is negative (minus $16), suggesting that the mean premium estimate from the social survey exceeded that from the verification. The bias was negative for all demographic groups and tended to be largest for younger families, the nonpoor, whites, and suburbanites. If the premium level as reported by the family was low ($76 or less), there was a positive bias; the bias was negative for higher values. This suggests that people reporting low premiums tend to underreport and those reporting high premiums tend to overreport.

We expected that a family's knowledge about its health insurance premiums might be influenced by the amount of medical care used. However, the field biases according to level of family expenditures for health services do not appear to vary systematically. We were also interested in knowledge of family premiums according to employer contributions to total premium. Table 4.2 suggests that when the policy is not obtained through a group, people have a better idea of their premium cost, as indicated by a relatively low reporting bias. The highest negative reporting bias occurs when respondents report that the employer pays none of the premium for a work group policy. The verification suggests considerably lower family premiums for these policies. Employer contributions not acknowledged by the respondent might be one reason for respondent overreporting of family premiums. In 55 percent of all group policies for which the respondent reported no employer contribution and we received adequate verifying information, the verification source indicated some employer contribution. This contribution averaged $161 per policy.

Services Covered by Health Insurance

The unit of analysis in this section is also the health insurance policy and the analysis is again limited to field bias. We will be concerned with whether each policy provides for inpatient hospital services, inpatient surgery and other physician services provided to hospitalized patients, major medical coverage, doctor visits, prescribed drugs, and dental care. Major medical coverage is defined as covering the range of most medical services received in the course of an episode of illness but requiring the patient to pay an initial deductible and some percentage of the cost thereafter. Policies supplementary to Medicare were included as basic hospital and/or surgical coverage but were not considered major medical policies. In computing the field bias for each type of coverage, the small number of cases where verification source did not know if a specific coverage was provided was treated as if the verification source indicated the coverage was not provided.

Table 4.3 compares the coverage for policies for which we have information from both the social survey and the verification. It shows that hospital, surgical-medical, major medical, and prescribed drug coverage are all more likely to be reported in the social survey than in the verification. The discrepancy seems to be particularly large for major medical coverage. In contrast, dental insurance is reported slightly more often in the verification than in the social survey.

Table 4.4 shows differences in biases according to various population characteristics for hospital coverage. In general the field bias is negative and considerably larger than the standard error. It tends to be largest for policies held by younger families, those with higher incomes, whites, and rural nonfarm residents. These groups were most likely to report that a policy covered hospital services when the verification indicated it did not. One reason for this bias is that families report major medical coverage as basic hospital coverage but we have not defined major medical policies as providing basic coverage unless they included a rider providing first-dollar coverage of hospital services.

Families who have not had any hospital admissions in the last year were, as might be expected, less knowledgeable about whether a policy covers hospital services. They tend to overestimate the extent of their hospital coverage. Enrollment method also influences knowledge about hospital coverage. Those policies not obtained through a work group but purchased directly were most likely to be described correctly with respect to hospital coverage.

Table 4.5 shows biases in reporting of surgical-medical coverage according to various population characteristics. These biases are also negative and generally larger than those for hospital coverage. Among the demographic subgroups, policies held by older families and poor families had the largest biases. This finding is in contrast to hospital coverage where the younger and higher-income families showed less bias.

As with hospital services, there appears to be a direct relationship between use of hospital services and knowledge of surgical-medical coverage. Also, people with work group insurance who pay their own premium appear better informed about surgical-medical coverage than those for whom the employer pays the total premium, as was the case with hospital coverage. However, those people who purchased their own policies, although best informed about hospital coverage, appear worst informed about surgical coverage.

The analysis of major medical coverage (Table 4.6) excludes policies described in either the social survey or the verification as a combination of major medical and basic hospital or surgical-medical coverage. This was done to make the definitions as similar as possible. Table 4.6 suggests that social survey respondents had a hazy understanding of whether their policies were major medical. Further, they erred in the direction of reporting coverage they did not have, as indicated by the negative sign of the bias. Among the demographic groups, older families and the poor had the largest biases, suggesting that they were particularly prone to claim major medical coverage that would later be denied by the verification source because policies supplementary to Medicare were generally not defined as major medical policies.

Somewhat surprisingly, there appeared to be little relationship between the major medical reporting bias and level of family expenditures for health services. Those who purchase their policies directly, as in the case of medical-surgical insurance, were most likely to claim major medical coverage that was not substantiated. Those with work group insurance who paid premiums themselves as well as those whose employer paid the total premium also tended to report more major medical coverage than was verified.

Table 4.7 looks at biases in reporting of insurance coverage of physician ambulatory care. (Major medical policies are not considered to provide physician visit coverage in this analysis.) Once again the overall bias is negative but there is considerable variance in the magnitude and sign of the bias. Certain groups, including older families, the poor, and farm and central city residents, actually have positive biases, suggesting they report less doctor visit coverage than was reported in the verification.

There was no consistent relationship between the total number of physician visits by the family and the reporting bias. Indeed, families with the most visits showed the largest negative bias. In this case people not enrolled through a work group and those enrolled in this way who paid their own premiums had little field bias. In contrast, when the employer contributed to the premiums, individuals were more likely to report doctor visit coverage that was not verified.

The overall bias in reporting of drug coverage is also negative. Although it is not large in absolute magnitude (−3.9%), it is considerably larger than the standard error and is half the estimate of drug coverage (Table 4.8), indicating considerable confusion about the provision of this option. As was the case for doctor visit coverage, the older families and the poor actually appear to report less drug coverage than was verified. There appeared to be little consistent relationship between total family expenditures for drugs and the field bias. Again, those with work group insurance paid for by the employer were least knowledgeable about their coverage, which they tend to overestimate.

The reporting of dental coverage (Table 4.9) differs from other types of coverage in that the overall bias is positive. The relative accuracy of reporting is fairly high compared to all other coverage except hospital coverage. Among the demographic subgroups, nonwhites significantly overreport dental care coverage. Poor families and those living in nonSMSA urban areas were most likely to underreport.

Neither level of expenditures for dental care nor enrollment status of policy appears to be systematically related to the field bias in reporting dental coverage.

Payment Sources for Inpatient Hospital Services

This section is concerned with the relative accuracy of various sources of payment reporting for hospital admissions. A processing bias as well as a field bias will be calculated because imputation was often used in deriving the estimates of total hospital expenditures (see Chapter Three). The unit of analysis is the hospital admission. Mean expenditures per admission are calculated for four types of payment: out-of-pocket, insurance, Medicare, and welfare. Key questions are (1) Is there differential bias in the social survey estimates by type of payer? and (2) Are these biases related to demographic characteristics of the population and overall level of hospital expenditures?

Table 4.10 shows that, in general, when payment information is provided by both the social survey and the verification, the total estimated cost is very close ($708 versus $703). Out-of-pocket estimates from the social survey are considerably higher than those from the verification. However, all major categories of third-party payers show higher estimates in the verification than in the social survey.

Table 4.10 shows the characteristics of admissions discovered in or rejected by verification to be very different from each other and also different from admissions for which we had information from both sources. The discovered and rejected groups have a relatively low estimate of both total and out-of-pocket expenditures. The rejected admissions had relatively high mean estimates for Medicare and welfare payments. Those discovered in verification had higher insurance estimates than did rejected admissions but a lower insurance estimate than did other kinds of admissions. Admissions for which the verification was incomplete had estimates from the social surveys that distinguished themselves from other types of admissions by the relatively high out-of-pocket estimates ($146). This category includes cases where the respondent refused to allow verification and where we had some indication that indemnity payments were made directly to the family. In the latter case the hospitals used as the verification source may tend to underestimate insurance payments because benefits might be paid directly to the respondent rather than to the hospital.

The next four tables provide estimates of the different kinds of biases for each of the payers listed in Table 4.10. These biases are reported by demographic characteristics of the patient's family and the level of expense for the hospital admission. (Table 3.2 combines the various payer biases into one set for the total admission expenditures. It is interesting to note that there is essentially no overall field bias, that is, the field biases for different types of payers go in different directions and tend to cancel each other.)

Table 4.11 shows that the social survey estimates for out-of-pocket expenditures for hospital admissions tend to be high for younger and higher-income families, whites, and suburbanites. Still worth noting is the finding that poor families had a mean out-of-pocket expenditure per admission of $90.

The field bias for out-of-pocket estimates for total admissions exceeded the standard error, suggesting a considerable amount of reporting error. The negative sign of this field bias indicates persons overreported their out-of-pocket expenditures. The overall processing bias is small relative to the standard error, for one reason because considerably less imputing was done for the out-of-pocket expenditures than for third-party payments. Respondents were more likely to report their own payments and less likely to provide information on third-party contributions. This processing bias is also negative, suggesting we tended to impute slightly higher amounts for the families' direct payment for in-hospital services than indicated by our verification sources.

Overreporting of out-of-pocket costs was considerably greater for younger families than for older ones and for white families than for nonwhite ones. However, nonwhite out-of-pocket mean expenditures were relatively low ($42), so that the field bias relative to the original estimate was higher for nonwhites than for whites. The absolute field bias was highest for suburbanites, but the out-of-pocket expenditures were also much higher for this group. Thus, the field bias relative to the estimate was actually lower for suburbanites' admissions than for those of most other residential groups. The absolute field bias for out-of-pocket expenditures also increased as admission expenditure levels increased. However, the field bias relative to magnitude of the estimate was not higher for the more costly admissions as compared to the less expensive ones.

The processing bias for out-of-pocket expenditures was relatively high for the lower-income families, rural nonfarm residents, and those with expensive admissions. For each group our imputation procedures tended to overestimate the magnitude of the bill.

Table 4.12 shows the biases for insurance payments for hospital admissions. The groups with the highest average payments according to the social survey estimates were the younger families, higher-income families, whites, suburbanites, and those with high expenditure admissions. The standard error is considerably higher than the field and processing biases for insurance payments. Also, the overall processing bias is larger than the field bias. Considerably more imputing was done for insurance contributions than for out-of-pocket payments because respondents were less likely to provide dollar amounts for insurance. The more imputing done for a given type of estimate, the greater the probability that there will be a large processing bias.

The largest positive field bias for insurance payments was for the nonwhite population, suggesting that nonwhite respondents underreported. However, the field bias was not uniformly positive among subgroups. The largest negative field bias was for the rural farm population. Respondents with the most expensive hospital admissions apparently slightly overreported insurance contributions.

The processing biases in insurance payment estimates were not consistent in sign among subgroups either. The groups for which the imputation procedure was most likely to underestimate insurance payments (indicated by a positive bias) were those with higher incomes, whites, and those living in central cities. In contrast, our imputation procedure apparently overestimated insurance contributions for admissions for nonwhites.

Table 4.13 shows field and processing biases considerably smaller than the standard error for Medicare contributions to hospital admissions costs. Both types of bias are positive, suggesting respondents tended to underreport Medicare contributions and the imputation procedure also underestimated them somewhat. By subgroup, positive field biases were largest for the poor, nonwhites, and central city residents.

Welfare payments, as expected, are highest for the poor, nonwhites, and central city residents, according to Table 4.14. The average overall field bias for welfare is positive, suggesting overreporting. The processing bias is negative, suggesting the imputation procedure tended to overestimate the welfare contribution. The magnitudes of both are somewhat less than the standard error.

Welfare includes both Medicaid and payments by local governments to hospitals caring for "free" patients.

The largest field bias among subgroups is for the suburbanites, the subgroup most likely to underreport welfare contributions and not to report admissions discovered in the verification process that involved welfare payments. In contrast, the poor, central city, and urban nonSMSA residents had sizable negative field biases. These groups apparently overreported welfare contributions and also had admissions they had reported as being paid for by welfare rejected in the verification process.

The processing bias is negative and relatively large for admissions of older families, the poor, nonwhites, rural farm and inner-city residents, and admissions with high expense. The bias relative to the size of the overall welfare contribution is particularly high for older families.

Tables 4.15 and 4.16 show how the differences among subgroups by various payers for hospital expenses are affected by bias adjustments. Table 4.15 suggests that the higher out-of-pocket expense for younger families compared to older ones is reduced considerably by the bias adjustments. The relative difference in out-of-pocket expenses between the most and least costly admissions is reduced also by considering the various biases.

The adjustments for insurance payments shown in Table 4.15 do not seem to result in any substantial changes in conclusions that might be drawn about subgroup differences. Similarly, Table 4.16 does not dramatically change the picture of subgroups differences regarding Medicare payments for hospital admissions.

The bias adjustments do appear, however, to influence the conclusions that might be drawn about welfare payments (Table 4.16). Differences between younger and older families appear to increase, with the latter having relatively larger welfare payments. The adjustments also result in considerable increases in differences between welfare payments for suburbanites compared to all other groups.

Summary

This chapter examined survey error in social survey estimates of family premiums paid for health insurance; the services covered by health insurance policies, including hospital, surgical-medical, doctor visit, major medical, prescribed drug, and dental; and various payment sources for hospital admissions, such as out-of-pocket, voluntary insurance, Medicare, and welfare.

A tendency was noted for families to report that they paid higher premiums than insurance companies and employers indicated the family paid. The reporting bias was negative for all demographic groups and tended to be largest for younger families, the nonpoor, and those with higher-level premiums. Amount of medical care used, contrary to expectations, did not appear to be related to the bias. However, the bias did vary according to how the policy was obtained and the extent of the employer's contribution in the case of work group insurance.

Hospital, surgical-medical, major medical, and prescribed drug coverage were all more likely to be reported in the social survey than in the verification. Respondents had special difficulty correctly identifying major medical coverage. Only dental coverage was reported slightly more often in the verification than in the social survey. The relationships of demographic characteristics, amount of use, and type of enrollment to reporting bias varied considerably according to type of coverage.

Examination of payment source for hospital admissions suggests that respondents report relatively higher out-of-pocket amounts but hospitals generally indicate relatively larger payments by third-party sources. These differences tend to balance each other so that estimates of total hospital expenditures per admission are quite close between the social survey and the verification. Major changes as a result of adjustment for biases among subgroups in the population were (1) reduction in the apparently higher out-of-pocket payments by younger families relative to older ones and (2) relatively higher welfare contribution to admissions of suburban patients than the social survey data alone indicated.

CHAPTER 5

Illness Conditions

Virginia S. Daughety

This chapter examines the error associated with survey individuals' (or their proxies') reports of the conditions for which they saw a doctor, the conditions for which they were hospitalized, and the operations they had during their hospital stay.

Social survey data (together with medical record data and clinical examination information) are a major source of information on morbidity prevalent in a population. Assessing the validity of reporting of conditions in an interview setting is important if social survey estimates of disease prevalence are to be used as reliable indicators of the health status of important population subgroups. Such indicators may be used to assess quality of life, to suggest need for additional health services, and to explain health and illness behavior. On a smaller scale the problem of reporting

Note: I express my appreciation to John Schneider, M.D., for his consultation regarding the diagnostic coding and matching procedures for this chapter. I also thank George Yates for his programming efforts and Ronald Andersen, Gretchen Fleming, Joanna Lion, and others for their helpful comments.

accuracy is relevant to the medical history commonly given by the patient at the start of a doctor-patient encounter (Mechanic and Newton, 1965) and for various health screening techniques that may or may not lead to contact with a practitioner.

Related Studies

Validity of social survey reporting of conditions has been previously examined by comparing the conditions reported in surveys of sample populations with those obtained in medical record checks or clinical examinations of the sample population (Elinson and Trussell, 1957; Mettzer and Hochstim, 1970; National Center for Health Statistics, 1965b, 1967, 1973b; Woolsey, Lawrence, and Balamuth, 1962). Conditions physicians or hospitals reported are the standard (true value) against which conditions respondents reported are compared. Discrepancies are attributed to respondent reporting error (field bias). The error estimates obtained from these studies are generally not net bias but estimates of "underreporting," that is, the percentage of conditions noted by the physician or hospital but not mentioned in the interview. Estimates of "overreporting," in other words, the percentage of conditions reported by the respondent that do not correspond to any provider-reported condition, are less frequent in the literature. This is partially because several major studies were designed around the availability of a single provider record source such as Kaiser Health Plan or Health Insurance Plan (HIP) and therefore did not attempt to contact other sources of respondent care. Other studies solicited morbidity information from the respondent whether or not a physician was seen for a particular condition. Hence generally these studies do not attempt to estimate "overreporting" or net field bias for respondents' condition reporting.

In the Elinson and Trussell (1957), Mettzer and Hochstim (1970), National Center for Health Statistics (1965b, 1967), and Woolsey, Lawrence, and Balamuth (1962) studies, which chiefly focused on chronic nonhospitalized illness, the percentage of underreported conditions ranges from 44 percent to more than 70 percent. One analysis of reporting error of physician visit conditions (those conditions for which a physician was seen) estimates an overreporting rate of 42 percent and notes that overreporting occurred to a lesser extent than underreporting, implying a positive net bias (National Center for Health Statistics, 1973b).

Available information on reporting error indicates considerably less underreporting for hospitalized conditions (those for which the respondent had a hospital stay) and for surgical procedures (operations) than for nonhospitalized conditions. Researchers' estimates of both hospitalized condition and surgical procedure underreporting range from 12 to 25 percent (National Center for Health Statistics, 1965a, 1965c; Richardson and Freeman, 1972). There is no available information on estimates of overreporting or of net bias.

There are two basic types of erroneous responses to survey questions about medical conditions: (1) respondent does not know the precise medical term and supplies another or says nothing or (2) respondent withholds true information and supplies another answer or says nothing. Theories set forth to explain these errors in physician visit and hospitalized condition reporting focus both on characteristics of the respondent and his or her conditions and on the nature of the interview itself.

Cannell and his colleagues explain error in hospital admission reporting in terms of memory and motivation (National Center for Health Statistics, 1965c). They suggest that length of hospital stay (a proxy for condition severity), self or proxy respondent, operations performed, and recency of the event determine how memorable the episode is to the respondent. Likewise, degrees of threat or embarrassment in reporting reasons for hospitalization and felt need for cooperation with the interviewer are given as factors affecting the motivation to report accurately. This scheme would appear to be applicable to *conditions* reporting as well. Cannell does not attempt specifically to analyze factors associated with hospitalized *condition* reporting error (although he does provide estimates of this reporting error). *Hospitalization* reporting can be expected to show less error overall because reporting the fact of admission is a necessary but not sufficient condition to reporting the condition involved.

Woolsey, Lawrence, and Balamuth (1962) suggest that reporting error for chronic conditions is related to the impact of the condition (as measured by volume and recency of services for that condition), the specificity of the condition and the ease of describing it in recognizable terms, and the respondent's understanding of what information was being requested. Mechanic and Newton (1965), in an overview of previous studies, affirm that salience of the condition (in terms of seriousness, volume of services, recency of services, and self-reporting) and degree of threat the condition represents are key to reporting error and suggest illness behavior of the individual as an additional factor. Madow (National Center for Health Statistics, 1967) emphasizes that communication between patient and physician is a relevant factor in understanding reporting error. Additional factors he mentions are the individual's health level, such impact measures as costliness of condition, and degree of disruption of routine the condition causes. However, he points out that although respondents report high-impact conditions with less error, the majority of respondent conditions are not particularly salient.

These researchers' findings generally tend to support the central idea of salience of conditions (in terms of self-reporting and impact) as an explanation for reporting error. For both hospitalization and physician visit condition underreporting, evidence from all studies indicates proxy respondents had a significantly higher degree of underreporting than self-respondents. Among proxy respondents alone, Cannell found an adult reporting hospitalization for a child was more accurate than an adult reporting for another adult (National Center for Health Statistics, 1965c). (Recall that hospitalization reporting can be expected to show less error than hospital condition reporting because reporting an admission is necessary but not sufficient to record the conditions involved.) Although Balamuth found proxy reporting for children and other adults considerably less accurate than self-reporting, he noted that proxy reporting for spouse was nearly equal to self-reporting for physician visit conditions (National Center for Health Statistics, 1965b).

Various condition impact measures showed a strong inverse relationship with reporting error—the greater the condition's impact, the less it was underreported. Number of physician visits for the specific condition and recency of these visits were strongly associated with lower rates of underreporting (National Center for

Health Statistics, 1967). However, *total* number of physician visits *for an individual* was not found to be related to accuracy of chronic conditions reporting (National Center for Health Statistics, 1965b, 1967). Cannell demonstrated that length of stay and recency of hospital admission were positively correlated with reporting of hospitalization; stays with surgery showed less underreporting of admission than those without (National Center for Health Statistics, 1965a). Other measures of impact of physician visit conditions, such as drug prescription, existence of pain or worry, and degree of disability for a specific condition, were positively related to the correct reporting of that condition (National Center for Health Statistics, 1967).

Unwillingness to report conditions representing a serious emotional threat or source of embarrassment was examined by Cannell as a reason for underreporting hospitalization (National Center for Health Statistics, 1965c). He identified disease categories such as diseases of male genital organs, psychoses, congenital malformations, malignant neoplasm of breast and genitourinary organs, and gonococcal infection and other venereal diseases as potentially threatening. He found a strong relationship between the high threat content of hospitalized condition or surgical procedures and a greater degree of hospitalization underreporting.

Several researchers also explored relationships between respondent cooperation and reporting accuracy. Both Balamuth (National Center for Health Statistics, 1965b) and Madow (National Center for Health Statistics, 1967) used the respondent's agreement to allow medical records review as a measure of cooperation. They found those withholding permission showed slightly to substantially higher rates of underreporting.

Madow (National Center for Health Statistics, 1967) found increased communication between physician and patient to be mildly related to lower underreporting rates for physician visit conditions and strongly related to how precisely the individual's report of his or her condition matched the physician's. Self-assessment of health was significantly associated with reporting error. He notes a marked difference in degree of underreporting of physician visit conditions for levels of perceived health, with those evaluating their own health as "excellent" (with fewer and less salient conditions) underreporting at twice the rate of those claiming fair and poor health.

The relationship of reporting error with sociodemographic characteristics such as sex, age, race, income, and education was assessed in nearly every study and was generally only weakly related to reporting error for physician visit conditions and hospitalizations.

For hospitalization reporting, Cannell found a significant difference in reporting error by race, with nonwhites being the poorer reporters (National Center for Health Statistics, 1965c). No significant relationship between race and reporting error was found for physician visit conditions. Low-income groups had greater underreporting of both hospitalizations and physician visit conditions (National Center for Health Statistics, 1965b, 1965c, 1967), although this was not found consistently. Elinson and Trussell (1957) found increasing underreporting with increasing income.

The relationship of reporting error with age was not conclusive. Elinson and Trussell (1957) found young adults fifteen to twenty-four and to a lesser degree persons sixty-five and over underreporting the fewest physician visit conditions; but Balamuth (National Center for Health Statistics, 1965b) found steadily decreasing underreporting as age increased. Slight differences in reporting of chronic conditions and hospitalizations by sex were found, although males and females are both claimed to have less underreporting in separate studies (National Center for Health Statistics, 1965b, 1965c, 1967). Cannell excluded pregnancies (one of the highly salient, easily dated, and therefore best reported conditions) in his analysis of hospitalization reporting and found that the difference between underreporting for males and females disappeared.

Methodology

This chapter develops and analyzes error estimates for three types of illness reporting occurring in the 1970 CHAS survey. The first type is physician visit conditions (any conditions for which the individual saw a physician in either an inpatient or ambulatory care setting). Included are visits for pregnancy, major (costly or chronic) illnesses, minor illness, hospitalization, and ophthalmological visits. The second type of illness is hospitalized conditions, that is, the main reason(s) the individual had a hospital stay(s). The third type is surgical procedures, that is, operations the individual underwent during a hospital stay, including deliveries and fracture reduction. Respondents were asked to name and provide information about conditions and surgical procedures they had in 1970 as well as the physicians and hospitals who treated them. Providers were subsequently contacted for similar information.

Three stages of data preparation were necessary in developing estimates of reporting error for these conditions and surgical procedures. These stages—coding, matching, and error computation—are explained in the next three sections.

Coding

Physician visits and hospitalized conditions respondents provided were coded into one of ninety-nine condition categories. These categories were adapted, with the assistance of a physician consultant, from the International Classification of Diseases codes (National Center for Health Statistics, 1957). Because ICDA codes were designed primarily for hospital discharges, new categories were added for conditions frequently occurring in primary care settings. For the coding of provider conditions, about one third of the ninety-nine categories were expanded into subcategories in order to preserve the greater specificity of physician and hospital reports. These subcategories were collapsed to permit rate and bias estimation for specific conditions in Tables 5.2 and 5.4.

Each surgical procedure reported by either respondents or providers or both was identified as one of forty-seven surgical procedure categories. The forty-seven categories were adapted from the most common operations reported by the National Center for Health Statistics. A supplement that lists and describes the ninety-nine diagnosis categories, the expanded categories, and the forty-seven surgical procedure categories is available from the Center for Health Administration Studies at the University of Chicago.

Matching

Matching is an attempt to determine the correspondence level between respondent and provider reports of conditions and surgical procedures. The unit of analysis for which matching was undertaken depended on the way the specific type of information was collected and processed. Physician visit conditions were matched on a per-person basis—all physician visit conditions reported for a person were matched with all physician reports of conditions (duplicate conditions were eliminated) for that person.

Hospitalized conditions were collected in the interview on a hospitalized illness episode basis rather than on a per-person basis. Conditions the respondent noted could refer to several hospital stays, all belonging to the same illness episode. However, provider reports of hospitalized conditions (in other words, discharge diagnoses) were collected on a per-stay basis. Therefore, provider stays were first grouped into hospitalized illnesses, and then provider reports of hospitalized conditions were matched with respondent reports for each hospital illness.

Surgical procedures were collected for each hospital stay from both respondents and providers. Therefore, surgical procedures were matched on a per-stay basis.

Besides the previously noted exclusions for duplicate provider conditions, certain other conditions and surgical procedures were excluded from the matching process. Exclusions were generally due to unavailability of provider information or unclassifiable reports. The complete set of exclusions is as follows (all figures are unweighted): For *physician visit conditions,* (1) 3,359 respondent conditions contributed by 2,311 persons remained unverified; (2) 489 respondent conditions (351 persons) were only partially verified and remained unmatched; (3) 116 respondent conditions (110 individuals) were unclassifiable; and (4) 73 respondent conditions (73 persons) had a provider report that failed to specify a condition (the corresponding respondent conditions are part of the second group). For *hospitalized conditions,* (5) 165 respondent conditions supplied by 125 individuals remained unverified; (6) 27 respondent conditions (27 persons) were unclassifiable; and (7) 33 respondent conditions (32 persons) had a provider report that failed to specify a condition (the corresponding respondent conditions are in the fifth group). For *surgical procedures,* (8) 84 respondent operations (77 individuals) remained unverified and (9) 14 respondent operations (13 persons) had a provider report that failed to specify a procedure (the corresponding social survey conditions are part of the eighth group). The total number of conditions and operations *included* in the analysis were (unweighted) physician visit conditions (social survey, 9,557; verification, 8,084), hospitalized conditions (social survey, 1,334; verification, 1,967); and surgical procedures (social survey, 571; verification, 573). Note that unverified/partially verified conditions (groups 1, 2, 5, and 8) *were* included in bias calculations for specific conditions and surgical procedures (Tables 5.2, 5.4, and 5.6).

The actual matching procedure was computerized. It was undertaken in three waves, one for each of the three levels of correspondence between respondent and provider reports; primary match (total correspondence), secondary match (some correspondence), and no match (no correspondence). Each provider and respondent condition (or surgical procedure) is matched at the highest possible level of correspondence.

The first wave examined conditions to identify all respondent conditions with perfect correspondence to conditions providers had reported and vice versa. In general, where both a respondent condition and provider condition have the same category code (among the ninety-nine condition codes or forty-seven surgical codes possible), these respondent and provider conditions are each permanently identified as having a "primary match." It indicates consensus between a respondent and providers that the respondent had a specific condition during the survey year.

The second wave compared the remaining respondent conditions with all provider conditions (already matched or not) in order to determine "secondary matches," that is, conditions with some degree of correspondence. The same thing is done with provider conditions remaining unmatched after the first wave. If an individual's condition is not exactly the same as a provider-mentioned condition but involves the same body system as or commonly occurs with the specific physician condition (as a symptom, cause, or result), a secondary match is granted. (These criteria were applied to all conditions by a physician consultant to determine which conditions and surgical procedures were associated as secondary matches.) A secondary matched condition indicates that the respondent was somewhat imprecise in describing a condition. This may be a sign that the respondent did not fully understand the nature of the condition or did not know the term to describe the condition and his or her vague response was coded into the wrong category. It also could reflect an individual's symptom, but the physician's report was the final diagnosis of a condition. In any case, a secondary match acknowledges common elements in a respondent's and provider's reports.

In the third and final pass all respondent conditions and provider conditions not granted primary or secondary matches are attributed "no match" status. "No match" respondent conditions, which are actually "overreports," could be due to deliberate misreporting but probably are vague descriptions of the condition or "telescoping" of a condition the respondent had some time before or after the survey year. Another possibility is that the physician made an error in recording the individual's condition or reporting it to the study. However, the physician's report should be less prone to error than the respondent's report; and for purposes of this study the physician report and those of other verification sources are assumed to be correct. Inaccuracies in physician record keeping have been documented (for example, American Public Health Association, 1969). These errors are probably more likely to be those of omission rather than commission and thus more likely reflected in "overreporting" rates. They also most likely involve less serious, more routine conditions or preventive treatment.

Unmatched provider conditions represent "underreports," that is, provider-reported conditions the respondent failed to mention. Reasons for respondent underreporting could include forgetting to report a condition, perhaps because it was insignificant, or remembering it inexactly. Embarrassment about mentioning a particular condition to an interviewer of the opposite sex could conceivably cause a respondent to supply a less-threatening condition or not respond at all.

Error Computation

Two different methods are used to estimate error in respondent (social survey) reports of conditions and surgical procedures. Field bias estimates for specific conditions and surgical procedures are computed in the same way as the biases presented in Chapters Two through Four (see Appendix A for the exact formula used). For field bias computations, primary and secondary matches are considered error-free answers and are treated identically.

Estimates of reporting error for population subgroups of interest are presented as summary accuracy scores. Instead of estimating net error by offsetting overreporting with underreporting, as is done in the field bias estimate, the accuracy score accounts for both sources of reporting error. The measure is constructed as follows:

$$\text{Accuracy Score} = \left(1 - \frac{a + b}{A + B}\right)100$$

where a = Number of "no match" respondent (social survey) conditions

b = Number of "no match" provider (verification) conditions

A = Total number of conditions the respondent reported in the social survey

B = Total number of conditions the provider reported in the verification

In the formula, primary and secondary matches are considered equally correct. (Unverified and partially verified-unmatched conditions are excluded from this computation.) A score of 0 percent indicates no correspondence between respondent and provider reports; 100 percent accuracy indicates neither underreporting nor overreporting by the respondent. The frequencies of primary, secondary, and no matches among all respondent conditions and provider conditions are also presented for population subgroups. These figures give an estimate of the precision of respondent reports and the relative contribution of overreporting versus underreporting to the summary accuracy score.

Findings

This section considers four types of personal and condition characteristics in order to describe and explain reporting error in physician visit conditions, hospitalized conditions, and surgical procedures: (1) individual demographic characteristics, including age, sex, race, income, and residence; (2) perceived health, including respondents' self-assessments of the state of their health and how much they worry about it; (3) the threat level of conditions and surgical procedures that may reduce the respondent's willingness to cooperate in the interview; and (4) measures of impact of the conditions in terms of volume and severity of conditions and use and expenditures of health services.

Conditions and surgical procedures are examined to determine relative levels of reporting error by the four characteristics. Overall accuracy, reporting precision (as measured by proportion of primary matches), and relative degree of over- and underreporting by the characteristics of subgroups are analyzed. Another consideration is field bias for specific conditions and surgical procedures. Throughout this chapter a finding referred to as "significant" or "strongly associated" is significant at the 95 percent level. (See footnote a, Table 5.1 for the test employed.)

Physician Visit Conditions

Physician visit conditions were reported with an overall accuracy of 56 percent (Table 5.1). Of the conditions reported by respondents, 42 percent did not correspond to any provider-reported condition, and 47 percent of provider conditions themselves remained unmatched. In terms of precision, 35 percent of all physician-mentioned conditions corresponded exactly (primary match) to social survey conditions.

Differences (Table 5.1) in reporting error of conditions for which a physician was seen by demographic characteristics are relatively small but significant. However, these trends vary somewhat from those reported in previous research. Reporting accuracy does not vary widely by age. Contrary to previous work, conditions for the older age groups appear to be reported with greater error and less precision—those sixty-five or older matched only 29 percent of their physicians' diagnoses with a primary match. Males are slightly better reporters than females and have lower rates of under- and overreporting. The largest differences occur by race and residence. Conditions reported by nonwhites show more reporting error (as in some previous studies), as evidenced by their 51 percent accuracy versus 56 percent for whites. Rural farm people appear to report their conditions somewhat less accurately than their urban counterparts.

The most surprising finding is the relative unimportance of self-reporting (grouped for convenience as a demographic variable) as an explanatory factor for reporting error. Accuracy scores for self-reporters (55 percent) and proxy reporters for adults (54 percent) are not significantly different, and proxy reporters for children have the highest accuracy (58 percent) and exhibit a considerably greater degree of precision than self-reporters or those reporting for other adults. One previous study successfully decreased the difference in error between self- and proxy reporting through the addition of more complete probes for illness (National Center for Health Statistics, 1965a). Similarly, the CHAS survey's focus on illness episodes for reporting of conditions and utilization may have narrowed the differences due to the respondent.

In contrast to previous findings, self-assessment of health level showed little relationship to accuracy; however, the conditions for those who worried a great deal about their health and obviously found health to be a salient issue were reported somewhat more accurately than conditions for individuals who worried less.

The measure of willingness to report conditions used here adapts a classification of threatening versus nonthreatening diseases used by Cannell to identify the conditions most likely to be misreported because they are threatening or embarrassing to relate (National Center for Health Statistics, 1965a). A higher rate of reporting error, decreased precision, and a higher rate of underreporting than overreporting would be expected for these conditions. These expectations are borne out in this study, where "very threatening" conditions have an overall accuracy six to eleven points lower (46 percent) than the less-threatening groups; and nearly two thirds of the threatening conditions doctors mentioned were not mentioned in the social survey.

Measures of the salience of an individual's conditions have been noted as among the strongest indicators of reporting error. The number of conditions for which the respondent claimed the individual saw a doctor is a summary measure of the impact of conditions on an individual's life. One would expect less reporting error in conditions by sicker people (as measured by more numerous conditions). This is clearly borne out—those reporting six or more conditions have an overall accuracy score of 73 percent (seventeen points above the mean) together with high precision and relatively low rates of over- and underreporting. Severity level of the condition is another impact indicator that displays a substantial negative relationship with reporting error. The most serious conditions, those for which a doctor "must" be seen, have an overall accuracy score of 62 percent. Also important is the higher rate of error for the least serious, "preventive care" category—it is apparently the result of a considerable degree of overreporting, indicating that over half the least serious social survey conditions were not matched with a physician report (perhaps attributable to their low salience level or relative inaccuracies in physician record keeping for minor conditions). This is coupled with an extremely low underreporting rate.

Other measures of the impact of illness on an individual, such as number of physician visits, out-of-pocket expenditures for these services, and expenditures for prescription drugs, would all be expected to be inversely correlated with error in conditions reporting. Aside from the obvious impact factor, the greater the number of visits an individual made to a doctor, the greater the opportunity for doctor-patient communications about conditions. Out-of-pocket expenditures would seem to be the measure of doctor visit expenditures most salient to an individual. Total expenditures for prescription drugs for illnesses also serves as an indicator of the summary impact of an individual's conditions. All three show the expected relationship, the strongest one being number of physician visits, which indicates a difference in accuracy score of fifteen points between conditions reported by persons with a single visit (48 percent) versus conditions for persons with fourteen or more visits (63 percent). (Recall that persons seen by more than one physician have a higher probability of attaining matches due to this study's matching algorithm for physician visit, which attempted matching social survey conditions with responses from any physician, whether or not the individual attributed that condition to a particular doctor.) A previous study showed no relationship between a person's overall frequency of physician visits and reporting accuracy (National Center for Health Statistics, 1965b, 1967). Differences in amount of an individual's out-of-pocket expenditures yields differences in the overall accuracy score of nine percentage points between low and high expenditures. High levels of prescription drug expenditures per individual are also associated with lower rates of reporting error.

Although unavailable in this study, information on visits and expenditures *per condition* would undoubtedly show an even more substantial relationship with reporting accuracy because all conditions are not equally salient to the individual.

In summary, the mean reporting accuracy score of physician visit conditions was 56 percent, comparable to previously reported levels of correspondence with verification reports. Although sex, race, and level of worry about health were strongly associated with accuracy, the expected differential between proxy and self-reporters was not present. A greater rate of reporting error was found for more threatening conditions. Salience of illness to an individual measured by several impact indicators was shown to have a strong negative association with reporting error.

Table 5.2 indicates field biases associated with conditions respondents frequently mentioned. Positive biases outnumber negative biases two to one, reaffirming that underreporting outstrips overreporting as a problem in condition reporting. As was expected, the threatening or somewhat threatening conditions in the table were underreported, with the largest positive biases associated with female trouble, chronic urinary or male genital conditions, and benign tumors.

Heaviest overreporting was noted for checkup and test and shot categories, two of the most frequently reported categories. The discrepancy between respondent and provider reports could be an indication in both cases of a physician recording specific conditions discovered or problems treated rather than the more general labels of checkup, tests, or injections. Conditions reported with the smallest net error (smaller than the standard error) include pregnancy/delivery, diabetes, and fractures and dislocations. Each is a precise, nonthreatening, and therefore well reported disease entity; so not only is the net error small, but in no case does the actual number of unmatched respondent or provider reports of these conditions total more than 25 percent of the matched reports. By contrast, the less precise categories—sprains, strains, whiplash, torn ligaments, and the common cold and various other acute upper respiratory conditions—all have a small net error that masks the fact that for every three matched respondent reports of strain or sprain there were at least two unmatched respondent reports and a like number of unmatched provider reports.

Hospitalized Conditions

We expected hospitalized conditions to be reported with less error than physician visit conditions, due to the more serious nature of conditions requiring hospitalization and the generally greater salience of hospitalized illnesses. The overall accuracy score of hospitalized conditions reporting was, indeed, significantly higher than that for physician visit conditions: 66 percent compared to 56 percent (Table 5.3). Underreporting accounted for more of the total bias than overreporting. Thirty-eight percent of verification diagnoses were unmatched and the same was true for 27 percent of social survey conditions. The overall accuracy score excludes all conditions for stays reported to be pregnancy-related in the social survey because conditions from these stays were reported with an extremely low rate of error in this study (as will be shown in a later section) and in previous research (National Center for Health Statistics, 1965c). Because these stays were a sizable proportion (over one sixth) of all stays reported in the social survey, the overall hospital accuracy score rises to 71 percent when pregnancy-related stays are included.

Some of the largest differences in error for hospitalized condition reporting are demographic differences, notably among categories of race, income, and residence. Race has by far the strongest association: Overall, nonwhite reporting shows a 56 percent accuracy score, significantly less than for whites (68 percent). The greater reporting error for nonwhites is due in large part to a high rate of overreporting, fifteen points above the mean for the sample—43 percent of respondent reports remained unmatched. This would seem to indicate a communication problem between nonwhite patients and their doctors due to cultural differences or perhaps continuity of care problems in public hospitals. Central city residents, a large percentage of whom are nonwhites, had the lowest accuracy score of all residential groups (59 percent versus about 70 percent for all other groups). They also exhibited high rates of underreporting. A fairly small though significant difference in reporting error was also shown by income.

Other demographic characteristics are apparently not associated with hospitalized condition reporting error. Differences by sex were almost nonexistent; both men and women exhibited similar patterns of precision and over- and underreporting, as well as the same overall accuracy score. Error was highest for oldest and youngest individuals, but these differences were not large.

Surprisingly, differences in accuracy between self- and proxy reporters are not evident, contrary to previous findings (but consistent with the findings of this study for physician conditions). The largest spread is five points difference between self-reporters and proxy respondents for adults and is not significant. Another interesting factor is the high degree of precision among respondents for children.

Only weak relationships with reporting error are apparent for measures like threat level of condition, self-perception of health, and number of conditions for which a physician was visited. This is especially surprising for the measure of number of conditions; Cannell (National Center for Health Statistics, 1965a) found it strongly related to admissions reporting accuracy. Level of threat actually seems to exert a slight trend in an unexpected direction—the most threatening conditions are actually reported with less error than less-threatening ones, but the difference is not significant. Over three fifths of "very threatening" verification diagnoses were primary matches, indicating a high degree of precision. This may be an artifact of variable construction because care was taken during categorization to label as very threatening only those diagnostic categories with a preponderance of threatening conditions, with the result that broader, less specific categories were rated as less threatening. Conditions in these less specific categories are, of course, harder to match.

Worry about health shows a strong relationship with reporting error—conditions reported by individuals with no worry have an overall accuracy score (77 percent) fifteen points higher than those who worry a great deal (62 percent). As anticipated, several measures of impact of the hospitalized condition, including total number of hospitalized days, occurrence of operation, and severity of condition, are associated with reporting error for the condition. Measures of recency of onset or treatment for hospitalized conditions (or for operations and physician visit conditions) were not available for this analysis.

Cannell mentioned number of days hospitalized as negatively correlated with error in admissions reporting. However, in this study conditions from illnesses resulting in a total number of days hospitalized of fewer than eight days are reported with significantly less error than those involving eight or more days. This discrepancy with previous research is partially because illnesses with multiple stays are most probably heavily represented in the eight or more days category; Cannell's research also showed that people with multiple admissions were less accurate reporters, as did our study (64 percent accuracy score). An illness involving an operation was expected, based on previous research, to be a highly salient event and thus reported with little error. This was borne out—conditions from illnesses involving operations were reported with significantly less error than those without operations (75 percent versus 60 percent accuracy score). Conditions involving operations are reported with a great deal of precision—57 percent of verification diagnoses were primary matches. This could be a result of the increased doctor-patient communication and patient awareness (due to gravity of the event and consent necessarily obtained) surrounding a stay involving an operation.

The relationship of severity to reporting error demonstrates the saliency argument from a slightly different angle. Hospitalized conditions that require only preventive care (checkups, diagnostic tests) or where physician care would make no difference (senility and some other chronic age-related conditions) were reported with a high rate of error (accuracy score of 27 percent). The fact that preventive diagnoses show the highest underreporting rate of all severity categories (56 percent) indicates the lack of impact of admissions involving these reasons. Most striking, however, is the rate of overreporting—four out of five least-severe conditions respondents mentioned remained unmatched.

A strong possibility is that although patients may think they have been admitted to the hospital for a "checkup" or "senility" (or may be unwilling to reveal a more serious underlying diagnosis, such as "checkup for cancer recurrence"), the underlying conditions are shown on the hospital records. Even if the admission is purely for a checkup or an evaluation of senility, other conditions may be shown because third-party payers rarely pay for preventive care on an inpatient basis.

Other conditions, ranging from symptomatic to very severe, are reported equally well. This seems to indicate that the saliency of an event as serious as an admission washes out the effect of severity of condition for all but the most mild reasons for hospitalization.

To summarize, such demographic factors as income, residence, and especially race showed the strongest relationships with reporting of hospitalized condition reporting. Proxy reporting unexpectedly was not associated at all. Persons who worried less had less reporting error. Measures of impact of illnesses, notably number of hospital days, number of admissions, and out-of-pocket expenditures for the illness, showed a weaker relationship with reporting error than expected—only the occurrence of an operation and severity level of diagnoses showed strong relationships with reporting error.

Frequently reported hospitalized conditions are listed in Table 5.4 with the associated field biases and standard errors. Once again underreports (positive biases) far outnumber overreports; however, only a few of the biases are larger than the standard error. Conditions with the largest biases include diabetes, heart trouble, hospitalized mental illness (each heavily underreported), and sprains, strains, whiplash, torn ligaments (overreported). As expected, pregnancy/delivery was the most accurately reported condition, with a field bias less than .5 percent of the social survey rate.

Surgical Procedures

In the previous section a strikingly low rate of reporting error was noted for conditions from hospital stays during which an operation occurred. This relationship between reporting error and occurrence of surgical procedure is examined in detail in this section. As Table 5.5 indicates, the mean overall accuracy score for surgical procedures (82 percent) is considerably higher than that for either hospitalized or physician visit conditions. Because virtually no deliveries are misclassified, women with deliveries are excluded from the table. Overall, 20 percent of all provider-reported surgical procedures remained unmatched. This was true for one of every six procedures respondents named. As would be expected with localized, highly specific conditions, surgical procedures were reported in the social survey with a great deal of precision—respondent reports were primary matches with provider information 72 percent of the time.

Clear relationships with reporting error are apparent for several demographic characteristics. Youngest children showed significantly less reporting error (93 percent) than the other age groups, who reported with average error. There was also a very low rate of both under- and overreporting for young children. For parents, then, an operation performed on their child is evidently a highly salient event.

Once again, no significant differences in male and female reporting can be found. However, procedures reporting for those with low family income and especially for nonwhites shows a significantly high level of error (76 percent and 61 percent accuracy scores respectively). Central city residents are worst operations reporters of all residential groups (71 percent); suburbanites have a high accuracy score (90 percent).

As with hospitalized conditions, error level by type of survey respondent showed only insignificant differences. Proxy reporters for children reported slightly better (86 percent) than self-reporters who had an average level of error (82 percent), although the difference is not significant. The overall error rate of proxy reporters for adults was highest but still only five points below the average reporting accuracy rate.

Perceived health variables did not appear to be particularly associated with reporting error. Even for highly threatening operations, no link with reporting error is evident. Surgical procedures are reported with more error for those who are seen to be in poorer health than for healthier individuals, although the differences are not great. Those who worried at all about their health were worse reporters (slightly below average) than those who expressed no worry at all (88 percent accuracy score); but again, significance is lacking. None of the available characteristics of the hospital stay during which the operation occurred, which were expected to serve as measures of salience of the event, shows any significant relationships with reporting error. Differences by length of stay, expenditures, or number of surgical procedures the individual underwent were minor.

Reporting error for operations is, on the whole, very low, unlike other forms of conditions reporting; and the error that exists is evidently not related to impact of characteristics of the operation or hospital stay or health characteristics of the sample person. The strongest association with reporting error is shown for low-income, nonwhite, and central city individuals, which seems to indicate a problem of patient-doctor communication possibly resulting from cultural differences and service utilization patterns.

Table 5.6 is a list of the most frequently reported surgical procedures and associated biases. Positive biases indicating underreporting are in the majority. The largest reporting biases are shown for fracture reduction (overreported), other operations on female genital organs, and skin operations (both underreported). The remaining conditions had field biases smaller than their standard error. Smallest field biases were associated with gall bladder (cholecystectomy) and cataract operations, resulting from nearly perfect respondent-provider correspondence.

Summary

The forgoing discussions of error in physician visit and hospitalized conditions and operations reporting indicate trends that vary at times from previously reported research. Table 5.7 lists factors found in this study to be significantly related to error for the three types of reporting considered.

Previous research had indicated mostly weak relationships between demographic variables and reporting error—inconclusive evidence of differences in accuracy by sex, better accuracy for older age groups, differences by race only for hospital admissions reporting, some differences by family income for physician and hospital reporting, and much better reporting for self- versus proxy respondents. This analysis has shown that nonwhite, low family income individuals, as well as central city residents (except for physician reporting), showed significantly greater reporting error. The finding most at variance with previous studies was that reporting accuracy did not vary as expected for self- versus proxy respondents (except for hospitalized conditions). For physician visit conditions reporting, in fact, proxy reporters for children showed consistently less error than self-respondents, with proxy reporters for adults showing the most error of the three.

A previous study showed that the high level of threat some conditions pose was associated with poor hospitalization reporting. In this study threat of condition was associated significantly (positively) with reporting error only for physician visit conditions.

Those who claimed to be in good health were shown in previous studies to be less accurate than those with poor health, as were those who did not worry about particular conditions they had. In this study degree of worry about health was a predictor of error for all but operations reporting—however, those who worried least were the best reporters for hospitalized conditions but the opposite was true for physician visit conditions.

Various measures of condition impact were found in earlier research to have a strong relationship with reporting error both for chronic conditions and for fact of hospital admission. Increased number and recency of physician visits for conditions, length and recency of hospital stays, occurrence of surgery for an admission, and drug prescription for a condition all were associated with decreased reporting error.

Our measures of impact of conditions on the individual, such as number and severity of conditions, number of visits or hospital days, costs, and occurrence of surgery, were related to error in physician visit and hospitalized condition reporting (strongest for physician visit conditions) but, curiously, not associated significantly with operation reporting. Perhaps the mere fact of having an operation is of such impact that the other measures become superfluous.

According to this study, then, demographic characteristics and impact measures seem to be the strongest predictors of reporting error for hospitalized and physician visit conditions, indicating that saliency of the condition (impact) and perhaps differing levels of physician-patient communication resulting from differing utilization patterns by demographic groups are operating together. For surgery, only demographic differences, probably reflecting differing levels of physician-patient communication, explained differences in reporting error.

CHAPTER 6

Problem Respondents

Judith Kasper

This chapter designates as "problem respondents" certain categories of respondents from the household survey and certain physicians contacted in the verification phase of the survey. The following discussion examines these persons' characteristics and their responses to this survey.

The first group of problem respondents is composed of persons who cooperated with the survey but refused to allow verification of their survey information. The absence of verification data made it impossible to evaluate the accuracy of the information they provided. (In our calculations of biases we attributed to this group the same degree of error that occurred for verified cases. It is quite conceivable, however, that this group is less accurate.)

Similarly, physicians who did not cooperate with our verification procedure hampered our evaluation of survey reporting. Comparing the characteristics of physicians who did not cooperate with those who did may provide insight into their different responses to the survey.

Two other types of survey respondents come under special scrutiny: those who gave inaccurate information (by under- or overreporting) and those adults who did not respond for themselves in the survey (those with proxy respondents). Each type of problem respondent will be discussed separately because they have little in common except the potential difficulties they cause in survey data analysis.

Persons Who Refused to Allow Verification

Following interview completion, we asked respondents to sign a permission form to allow us to contact physicians, hospitals, and insurance companies in the verification phase of the study. We requested permission of families to verify social survey data of families and it was usually given or refused for all family members. In a few cases, however, permission was granted for some family members and not for others or permission was granted to contact some physicians but not others. A small number of respondents (12.2 percent) refused permission.

Two explanations for people refusing permission to verify survey data are suggested. One is that refusing verification is part of a general pattern of respondent uncooperativeness, another symptom of which would be incomplete or inaccurate reporting of information. Such uncooperativeness may stem from suspicion of surveys in general or a concern for personal privacy. A study of morbidity from chronic conditions (National Center for Health Statistics, 1965b) using respondent data and medical record information obtained a rate of permission similar to ours to verify reported data (89 percent agreed). The authors suggested completeness of reporting might be related to willingness to give permission but found only a slight correspondence between the two.

A second explanation for refusal to allow verification is that health is not a salient issue for some persons and their interest in the survey may not be sufficient to have them cooperate beyond the interview. There is evidence that individuals are more likely to cooperate with both clinical and interview surveys about health if they see some potential personal gain or derive satisfaction from discussing health (National Center for Health Statistics, 1968, 1969).

One group that might be expected to share some characteristics with persons who would not allow verification was nonrespondents, by definition the most uncooperative persons in the sample. Persons who refused to allow verification became uncooperative at the verification stage of the survey. Both nonrespondents and persons who refused to allow verification were more likely to have incomes above the poverty level. (Table 1.1 shows nonrespondents' characteristics.) There was little difference in response rates by race; but once nonwhites agreed to be survey respondents, they were somewhat more likely than whites to allow verification. (However, the difference is not significant at the two-standard-error level; see Table 6.1.) Urban residents (SMSA central city and SMSA other urban in Table 6.1) were both more likely to be nonrespondents (Table 1.1) and more likely to refuse permission to verify if they agreed to be survey respondents. Although there are some similarities between the two groups, the evidence does not support any firm conclusions about the relationship between nonrespondents and those who refused to allow verification following cooperation with the household survey.

The cooperation variables shown in Table 6.1 also addressed the general cooperativeness of persons who refused permission to verify. These variables examine completeness of reporting and two measures of cooperation on interviewer calls. Persons who would not report their income in the survey were much less likely to have granted permission to verify (36.4 percent refused permission compared to 11.2 percent for persons who reported their income). A Survey Research Center study of consumer economics found that most people know their family income; their willingness to report it depends largely on their motivation (National Center for Health Statistics, 1965a). Unwillingness to report income and refusal to allow verification appear to have something in common—the respondent's decision to conceal certain information or at least to refrain from complete cooperation with the survey.

Failing to report physician or hospital expenditures also might be construed as indicators of an uncooperative respondent and one who would be less likely to give permission to verify data. Whether the respondent provided information about physician and hospital expenditures showed no relationship to permission to verify, however, suggesting that incomplete reporting of physician and hospital expenditures is probably due to lack of knowledge about these expenditures rather than to lack of respondent cooperation. Persons who were cooperative or completed the interview in one or two calls were also less likely to have refused verification (98.1 percent for persons who required only one call versus 22.9 percent for those who had five calls or more). "Cooperation on calls by interviewer" is a measure of the outcome of each call in which the interviewer attempted to contact the respondent (for example, interview completed, delay but respondent cooperative). The categories of the variable are shown in Appendix C. For the purposes of this chapter the variable was dichotomized into "cooperative" (1–3) and "uncooperative" (4–7) rather than the division shown in the definition.

Persons who worried or had more severe diagnoses were more likely to grant permission to verify survey data (Table 6.1). Persons with four or more diagnoses were most likely to give permission (96.5 percent did so), but persons with no diagnoses were significantly more likely than those with one diagnosis to allow verification. (Joanna Lion, the original study director of this project, suggested that people who have no diagnoses, and therefore no medical care, may be willing to give permission precisely because there is little or nothing to verify.) These findings support the proposition that individuals are more cooperative, which includes giving permission for verification, if they are interested in health or health is a salient issue for them (Mechanic and Newton, 1965).

It is difficult to make inferences about the accuracy of the survey information provided by persons who would not allow verification. However, persons who refuse to report their income and have lower overall cooperation might be more likely to conceal or misreport information. Although persons who worry less or report less severe diagnoses are also more likely to refuse permission, there is no reason to expect these respondents to be less accurate than others.

Physician and Clinic Cooperation with the Verification

Even when respondents gave permission to contact their doctors or clinics where they received care, these providers did not always cooperate with the verification effort. (Doctors or clinics we were unable to locate, physicians who were deceased or retired, and physicians of persons who refused to allow verification were excluded from this analysis. Also excluded were cases in which verification was attempted but the information obtained was inadequate. Cases in which the verification form was never returned as well as outright refusals were counted as not cooperating.) Two variables that might influence cooperation were examined: physicians' characteristics and the number of times a physician or clinic was asked to verify respondent information.

It was expected that older doctors might be less likely to cooperate, both because of the effort involved and because they might be more conservative in releasing patient information. Specialists and board certified physicians were expected to be more likely to cooperate, being better educated and more familiar with research. However, Table 6.2 shows that physician cooperation did not differ greatly by age or whether they were board certified. Specialists other than primary care specialists were more likely than general practitioners to cooperate, but there was little difference between general practitioners and either general surgeons or doctors with primary care specialties.

Another element that might influence cooperation is the amount of effort involved, one measure of which is the number of patients for whom the provider was asked to verify data. For each patient the provider was asked to complete the front and back of a single-page form. As Table 6.3 indicates, some physicians and clinics received twenty-one or more verification forms. A few clinics, in fact, received over one hundred forms each. Although 91 percent of the physicians and 93 percent of the clinics complied when asked to verify the experience of only one patient, partial cooperation (for some patients but not others) and noncooperation increased with increasing numbers of verification requests. The sheer number of patients for whom a physician or clinic was asked to provide verification appeared to be a much more important factor in cooperation with verification than physician characteristics.

Inaccurate Respondents

Concern over accuracy of respondent reporting is the primary justification for doing a verification of survey information. Chapters Two and Three discussed field bias or respondent reporting error. In Tables 6.4 and 6.5 this error is divided into error due to underreporting and error due to overreporting. This division allows us to look at the independent effects of these two types of error; the combined effect is shown in Chapters Two and Three. For groups where these two types of error occur with the same frequency, the overall field bias will appear to be negligible. In other words, if overreporting and underreporting cancel each other out, the effect on the total mean estimate is insignificant. Therefore, to evaluate how accurately respondents are reporting information, over- and underreporting must be examined separately.

This section focuses on persons who underreported or overreported five types of expenditures or service use—hospital admissions, number of hospital days, expenditures for admissions, number of physician visits, and expenditures for physician office visits. In each case the analysis is restricted to reported social survey information for which we received a response from a verification source (data imputed by the office staff are excluded). Persons received an accuracy score based on comparing social survey with verification data. The formula for comparing reported social survey and verification data to differentiate under- and overreporting was

$$\frac{\text{Social Survey} - \text{Verification}}{\text{Social Survey} + \text{Verification}} = \text{Accuracy Score}$$

A negative result indicated underreporting; a positive result indicated overreporting. Accurate reporters were those whose social survey report was within 25 percent of verification. For underreporters verification exceeded social survey by greater than 25 percent. For overreporters social survey exceeded verification by greater than 25 percent.

Underreporting is generally attributed to memory or motivation problems. It may stem from a long recall period (Neter and Waksberg, 1964) or infrequent and uneventful, thus easily forgotten, contact with the health care system (National Center for Health Statistics, 1965b, 1965c). Persons may also underreport because they are withholding information they consider threatening or embarrassing in some way (National Center for Health Statistics, 1965a). Previous studies suggest certain types of persons are more likely than others to underreport information—persons in low-income families, older persons, nonwhites, and persons with less education (National Center for Health Statistics, 1965a, 1965c). Proxies who report for other family members tend to underreport as well. The number of physician visits for a condition, the length of a hospital stay, and other factors that increase or decrease the eventfulness of receiving care also were related to underreporting in these studies. Although our study is less reliable in judging underreporting, we are able to detect underreporting by persons who reported some use of services.

In many studies overreporting is not considered as serious a problem as underreporting. Researchers tend to be more concerned about information respondents fail to report than about the reporting of events that did not occur. Most of the previously mentioned studies of underreporting proceeded from patient records to patient interviews and focused on what was not being reported; thus they are unsuited for evaluating overreporting. Our study, which is a less reliable guide to underreporting, is a good source of information about overreporting because we first interviewed a sample population and then attempted to verify reported contacts with the health care system. One important cause of overreporting is an effect called "telescoping," the respondent's tendency to move events to earlier or later time periods (Neter and Waksberg, 1964; Sudman and Bradburn, 1974). This type of overreporting is almost certain to occur in a survey that asks respondents to remember events over a one-year period. Events such as hospitalizations near the beginning or shortly after the end of the recall period may be reported as having occurred during the survey year. Another study, which examined mothers' reports of clinic visits, suggested that overreporting increases with increasing numbers of visits (Kosa, Alpert, and Haggerty, 1967).

Accuracy in reporting hospital admissions, length of stay, and expenditures for admissions was generally high (Table 6.4). Eighty-eight percent of those reporting hospital admissions accurately reported admissions occurring during the survey year. Approximately 82 percent of those reporting admissions were able to report length of stay and related expenditures accurately. (In Tables 6.4, 6.5, and 6.6, the characteristics shown are those of the designated respondent rather than the person who may have responded for him or her.)

The oldest and youngest age groups were the least accurate reporters of hospital admissions, length of stay, and expenditures. Both groups tended to overreport the number of days spent in the hospital and underreport expenditures. Because persons under seventeen rarely reported for themselves, their inaccuracy is the proxy reporters' fault. However, the oldest age group had a large percentage of self-respondents (see Table 6.6). Low-income persons and nonwhites were generally less accurate reporters. Once again both groups were more likely to overreport number of days and considerably underreport expenditures (20 percent of the low-income and nonwhite did so). Central city residents were also less accurate reporters, but the greatest difference in reporting accuracy by residence was the consistently more accurate reporting by urban residents outside the central city. Persons with less education were more likely to be inaccurate reporters; those with the most education were most accurate. Proxy respondents for adults and self-respondents had similar accuracy levels, but proxy respondents for children were less accurate reporters of days and expenditures (overreporting one and underreporting the other, as previously mentioned).

Differences in reporting accuracy were not as evident for the perceived and evaluated health variables. Persons who perceived their health as excellent or who did not worry were somewhat more accurate, but so were persons with higher numbers of diagnoses. Individuals who were uncooperative were most strikingly inaccurate in reporting hospital admission expenditures, a similar number underreporting (14.8 percent) and overreporting (15.6 percent). Persons whom the interviewer had to contact at least five times exhibited a similar pattern but more often tended to overreport days and expenditures. Lower expenditures for admissions were associated with less accurate reporting, particularly of expenditures. Underreporting of days and expenditures was more common for the low-expenditure group, supporting the notion that persons with low levels of use are more likely to underreport. Persons with high expenditures were more likely to overreport expenditures, leaving the middle group the most accurate.

Table 6.5 examined accuracy of reporting physician visits and physician office visit expenditures. A visit to a doctor has less impact than a hospital admission, and the lower overall accuracy in reporting physician information reflects this. (Of those reporting physician contact, 53 percent accurately reported their visits; 45 percent were able to report expenditures for office visits accurately.) Low-income persons and nonwhites were less accurate reporters of visits and tended to overreport number of visits during the year. Central city residents were less accurate than other urban residents. Although differences in reporting visits by educational level were small, persons with the least education were more likely to overreport and persons with the most education were more accurate than others. Differences in reporting office visit expenditures by other demographic characteristics were small.

Perception of health and worry about health affected reporting of both visits and office visit expenditures. Persons who perceived their health as excellent or good were more likely to underreport visits and expenditures; those who perceived their health as poor tended to overreport. Similarly, persons who did not worry about their health underreported their physician visits and persons who worried a great deal overreported visits and expenditures. Surprisingly, persons who were uncooperative or required five or more interviewer calls were at least as accurate as more cooperative individuals (this was not the case with hospital admission information).

The greatest differences in reporting accuracy for physician visits and office expenditures occurred for the level of use variables. Persons who reported one visit on the social survey underreported their visits but those with five or more visits overreported. The same pattern held for the two expenditure variables. Here, persons with low levels of expenditures underreported but those with high levels overreported. Similarly, persons with one visit or low levels of office visit expenditures and those with many visits and high levels of expenditures overreported office visit expenditures.

As previous studies have found, old persons, the poor, the uneducated, and nonwhites were less accurate reporters of hospital admissions, length of stay, and expenditures (although differences in reporting physician data by demographic characteristics were small). Perception of and worry about health affected reporting accuracy for physician visits and office expenditures but not for hospital admission information. Level of use was also an important factor in accurately reporting physician information.

Physician visits and office visit expenditures are not particularly salient events in the course of a year. Consequently, variables related to memory and salience—levels of use, whether individuals worried about their health, whether they perceived their health as good or poor—showed the most differences in reporting accuracy for physician data. Hospital admissions, however, are salient. Because people were unlikely to forget an admission, other factors—personal characteristics (age, education, and so on) and motivation (whether the person was cooperative, the number of interviewer calls involved)—appeared to be important in misreporting.

Self- and Proxy Respondents

Most surveys find information individuals provide about themselves more reliable than information one individual provides about others in the family. (Tables 6.4 and 6.5 showed this to be true for our survey as well, but the differences were not as great as might be expected.) In the CHAS-NORC survey, which limited each household to two main respondents, 38.3 percent of the sample responded for themselves and 67.7 percent had proxies responding for them. Children (sixteen and under) did not respond for themselves; they were 33.9 percent of the total population.

Certain individuals in a family were more likely than others to be self-respondents. Table 6.6, which is restricted to adults, showed that older persons were more likely to be self-respondents than any other group (75.2 percent of persons over sixty-five were self-respondents). Also, more poor persons were self-respondents. Women were considerably more likely to respond for themselves than males, even when the women were employed fulltime.

Adults who perceived their health to be excellent or did not worry at all about it were significantly less likely than others to be self-respondents. Similarly, persons with nonsevere diagnoses were less likely to be self-respondents than those with severe diagnoses. And persons with one or two diagnoses were much less likely to report for themselves than persons with three or more diagnoses.

As mentioned, Tables 6.4 and 6.5 showed self-respondents were only slightly more accurate in our survey than proxy respondents. Table 6.6 indicated self-respondents in our survey were more likely to be old, have a low income, and be in poorer health. These groups were frequently less accurate reporters (Tables 6.4 and 6.5) and may have depressed the accuracy of self-respondents in relation to proxy respondents for adults.

Summary

This chapter examined several types of "problem respondents" of concern in the CHAS-NORC survey. It was suggested that persons who refused to allow verification did so for two reasons. Refusal to allow verification was part of a general pattern of uncooperativeness that included an uncooperative response to interviewer calls and refusal to report income in the survey. Second, refusal to allow verification reflected a lack of interest in health by some respondents who were in good health and had infrequent contacts with the health care system. Both propositions received support from the data.

In this chapter we examined two variables that might affect physician and clinic cooperation with verification—physician characteristics and the number of times a physician or clinic was asked to verify respondent information. Physician characteristics appeared to have little influence on cooperation; but the number of patients for whom a physician or clinic was asked to provide verification data was an important factor. Cooperation dropped from around 90 percent cooperating fully for one patient to around 50 percent cooperating fully for twenty-one or more patients.

Reporting accuracy was generally high for hospital admissions, length of stay, and expenditures (between 80 and 90 percent of the respondents were accurate). Seeing a doctor is an event with less impact than a hospital admission, and the lower reporting accuracy for physician visits and office visit expenditures reflected this (53 and 45 percent of respondents were accurate, respectively). The high accuracy level of reported hospital admission information suggests this information is easily recalled. Inaccurate reporting of hospital days and expenditures, in particular, appeared to be related to personal characteristics (age, education, and so on) and motivation (cooperativeness, number of calls required). Misreporting physician information appeared to be related to variables that might influence memory—levels of use (number of visits, high or low expenditures), whether persons worried about their health, whether they perceived their health as good or poor.

Most surveys find information provided by self-respondents more reliable than that provided by proxy respondents. In the CHAS-NORC survey each household was limited to two main respondents and children did not respond for themselves. Persons sixty-six and over, the poor, and females, whether employed full time or not, were most likely to be self-respondents in our survey. Those in poor health were also more likely to be self-respondents. Self-respondents were only slightly more accurate in reporting physician and hospital admission information than proxy respondents for other adults, perhaps because the persons most likely to be self-respondents in our survey (the old, low income, those in poor health) were also associated with less accurate reporting.

CHAPTER 7

Adjusting for Total Nonresponse

Martha J. Banks
Martin R. Frankel

✤ ✤ ✤ ✤ ✤ ✤ ✤ ✤ ✤ ✤ ✤ ✤ ✤ ✤ ✤ ✤ ✤

In almost any survey there will be cases designated for interview that were not actually interviewed. Some potential respondents may have refused to be interviewed; a few initially began the interview process but broke it off before completion; others were not at home when repeated interview attempts were made; some were away from home during the entire interview period; and a few were not interviewed for various other reasons, such as a language barrier. Making no adjustment for nonresponse implicitly assumes that nonrespondents do not differ from respondents in any characteristic of interest. The degree to which they *do* differ is proportional to the amount of bias introduced by ignoring nonrespondents.

Basic Features of Nonresponse Adjustment

Any approach to nonresponse adjustment consists of two elements. First, the population must be categorized into subgroups and the response rate for each group must be determined. The categories chosen to form the subgroups not only should be correlated with characteristics of interest in the study (for example, mean number of dental visits) but also must be determined without having to obtain the information from the potential respondents themselves. Dwelling location (for example, in the central city of an SMSA), type (duplex, and so on), and state of repair can be observed; such characteristics as race and presence of children can be determined from neighbors; and number of calls made and their results can be obtained as part of the interviewer control and record-keeping process.

The second major element in nonresponse adjustment is that of determining the values to impute to the nonrespondents. It is usually simplest and not unreasonable to assume that respondents and nonrespondents falling into a given category have identical characteristics. That is, nonrespondents living in well-kept housing in central cities have about the same health characteristics as do their responding counterparts; nonresponding and responding persons living in dilapidated central city housing are very similar; and so on. This method (and the rationale for nonresponse adjustment) is summarized in National Center for Health Statistics (1971, p. 22): "The adjustment for nonresponse is intended to minimize the impact of nonresponse on final estimates by imputing to nonrespondents the characteristics of 'similar' respondents, that is, by relating respondents to nonrespondents by ancillary data known for both."

Sometimes, however, we have evidence that respondents and nonrespondents in the same category do *not* have nearly identical characteristics. This evidence may come from external sources or from the current study itself (by choosing a sample of nonrespondents, vigorously attempting interviews with those chosen, and weighting and extrapolating from the results to obtain a final estimate).

For example, if initially there was a 65 percent response rate for a category that produced a mean value of 105.52 for some characteristic and if one fourth of the remaining 35 percent received more interview attempts, resulting in a 45 percent response rate and a mean of 110.63, extrapolation for this particular characteristic might lead to an estimated mean of 112.84 for the remaining 55 percent. Thus, the overall estimate would be $.65(105.52) + .35\,[.45(110.63) + .55(112.84)] = 107.73$.

This would be a rather expensive method if it were implemented after the interviewers had already attained a fairly high completion rate, especially if most of the remaining nonrespondents were refusals rather than never-at-home's because few of them would be expected to change their minds about being interviewed. But there are problems with implementing this method by subsampling respondents before obtaining a high completion rate. Interviewers either may be unwilling to give up on cases before they have done everything they can to get an interview or, conversely, they may reduce their level of effort because they know they may be able to drop some cases after a few interview attempts. Thus this method should be used only when interviewers' work is carefully controlled and monitored.

Another administrative problem with this method concerns the computation of a completion rate. Subsampling initial refusers and never-at-home's would result in a lower completion rate than would the usual approach, but the data would be more representative of the surveyed population, because it would include more of the hard-to-interview people. Some kind of "effective completion rate" might need to be devised in order to reflect more nearly the true representativeness of the sample data. The effective completion rate would lie between the actual completion rate and a rate based on the subsample of initial nonrespondents.

Compared with sampling nonrespondents, it is easier and less expensive to use external data—data from a previous study that either had used the above approach or had access to the administrative records of both survey respondents and nonrespondents. Care must be taken in adjusting previous data according to current patterns, including nonresponse patterns.

Recent sampling literature has addressed the question of nonresponse adjustment. Warwick and Lininger (1975) devote a section to missing data, data missing chiefly due to nonresponse. Moser and Kalton (1972) provide nonresponse rates from a number of surveys and discuss several adjustment methods. Daniel (1975) expands on these methods. Earlier, Stephan and McCarthy (1958) dealt with accessibility and cooperation of prospective respondents. More technical works include Kish (1965), Cochran (1963, 1978), and Hansen, Hurwitz, and Madow (1953).

Our Approach

Data from the 1970 CHAS study were adjusted for differential nonresponse by subsample. (These subsamples are described in Chapter Nine.) This adjustment assures that the data have the same relative number of interviewed households by subsample as was the case for the assigned households and was done because of the positive correlation between subsample designation and most characteristics of interest in the study. The subsamples grouped geographic areas and persons within them by percentage of elderly, poor, or both, characteristics that influence health and medical care use.

The estimates of nonresponse bias appearing in previous chapters were obtained by first using screening information to estimate response rates by independent variable categories and then using external data to estimate how the dependent variable values differed between nonrespondents and respondents. The estimated response rates were based on data that not only were adjusted by subsample but also contained a post-stratification adjustment. (This adjustment is discussed in Chapter Nine). Thus there was an implicit adjustment for differential nonresponse among post-stratification categories.

Alternative Approaches

With the data we had available, we chose to try two approaches as alternatives to that described in the previous section. Both assume that respondents and nonrespondents within the same category tend to have the same values. Thus the overall results of the two methods differ only because of different choices of categories. The first alternative creates categories based on geographic location; the second uses information about the reason no interview was obtained.

The geographic adjustment method used primary sampling unit (PSU) and sample designation (see Chapter Nine) as the category determinants. Using sample designation effectively grouped cases within PSUs by presence of the poor, the elderly, or both. This method is similar to that used by the Census Bureau in many of its surveys (Thompson and Shapiro, 1973, p. 114).

The other method used categories based upon the reason no interview was completed. To adjust for cases who refused to be interviewed or who broke off in the middle of the interviewing process, we increased the weights of those respondents who were not completely cooperative. "Uncooperative" respondents were those who broke appointments with the interviewer, refused to give permission to verify, and so on. In this study the majority of nonrespondents were refusals. To adjust for other cases, those never found at home or unavailable for the entire field period, we increased the weights of all cases according to number of calls needed to complete the interviews. We chose to assume that it would have taken 20 percent more calls to obtain interviews from the never-at-home's. (For example, if there were ten households where no one was ever at home after 7 calls, we estimated it would take an average of 1.4 more calls to obtain interviews. Six would take 1 more call, and four would take 2 more.) Kish (1965, p. 537) discusses differentiating between refusals and never-at-home's.

Table 7.1 shows the factors used to adjust for noninterview by reason. A similar table is not presented for geographic adjustment because it would contain about five hundred entries. In computing the factors to be used for noninterview adjustment, we used an upper bound in order to control the amount of differential weighting caused by nonresponse adjustment. The United States Bureau of the Census (1978, p. 59) uses an upper bound in its estimation procedures. The upper bound we have chosen is 2.3995, which is twice the overall average adjustment, that is, twice the number of assigned, eligible households divided by the number of interviewed households. When a calculated factor exceeded this bound, we reduced it to the largest acceptable figure (2.3995) and increased the factor of a similar category appropriately. (See Appendix B for an example.)

Table 7.2 compares the percentages of interviewed households that received various levels of noninterview adjustment weights according to each method. Most of the sample households were given weights between 1.02 and 2.00 by the geographic method; most received a weight outside this range (1.02 or less or more than 2.00) from the adjustment based on the reason that no interview was obtained.

Tables 7.3 through 7.7 attempt to assess the impact of each of the three noninterview adjustment methods. We present the results that we obtained from their use without intending to imply what the expected results would be if these methods were applied to a number of similar data sets.

Table 7.3 provides information about the effect of noninterview adjustment on the distribution of sample persons. The column labeled "adjusted using external data" was obtained by multiplying the unadjusted social survey estimates by the overall family response rate and dividing by the family response rate for the characteristic. These rates appear in Table 1.1. Because they are family response rates, this column in Table 7.3 implicitly assumes that the mean number of persons per household is the same for interviewed and noninterviewed households of the same type. The estimated percentages of persons with adjustment using geographic information and using information about the reason no interview was obtained are the result of applying noninterview adjustment factors to all persons in a given category of households.

All differences shown in the table are very small; however, the population estimates adjusted using geographic data are more similar to the unadjusted estimates than are those adjusted using the reason for noninterview. The latter increased the proportion of the sample who were above poverty, were white, lived in small towns, never worried about their health, had only elective diagnoses, had one or two diagnoses, saw a physician once, and were not admitted to a hospital during the year. It reduced the suburban population, those in fair health, those who worried a great deal about their health, those with all mandatory diagnoses, those with three or four diagnoses, those who did not see a physician, and those who had exactly one hospital admission during the year. The estimates adjusted using geographic information decreased the central city figures and increased the suburban figures.

Tables 7.4 and 7.5 show the effect of nonresponse adjustment on two different estimates—mean number of physician visits for persons seeing a physician and mean hospital expenditures per person. Both tables show the adjustment based on geographic information has less effect on the means than does the adjustment based on the reason that no interview was obtained. The nonwhite mean hospital expense per admission is clearly the most volatile. The estimate varies by as much as $100, depending on the nonresponse adjustment method. In general, the difference between the unadjusted estimates and the estimates adjusted using external data is greater than the difference between the unadjusted figures and those adjusted using geographic information or the difference between the original figures and those adjusted based on reason for nonresponse. This is expected because the external data did not assume that respondents and nonrespondents in the same adjustment category have identical health experiences, but the other adjustments used this assumption. The noninterview adjustment seems to have had more effect on hospital expenditures than on physician visits, at least for totals and among the demographic characteristics.

The effect of nonresponse adjustment on differences of mean physician visits and of mean hospital cost between population groups appears in Tables 7.6 and 7.7. Following each difference is the ratio of difference to the estimated standard error of the difference. Table 7.6 shows little or no noninterview effect on differences in mean visits. All differences except those by race and age are highly significant statistically.

Because there were many fewer instances of hospital admissions in the sample than of persons seeing a physician, the statistical significance of the differences appearing in Table 7.7 would be expected to be less than in Table 7.6, even if the population represented by the sample differed equally in these two health characteristics. So either for this reason or because there really is less difference in hospital expenditures between groups than there is difference in physician visits or partially for both reasons, the significance ratios tend to be smaller in Table 7.7 than in Table 7.6.

Among the demographic variables, the differences seem to be larger in Table 7.7 for the estimates adjusted based on noninterview reason, especially differences based on family income and on race. In the health categories the differences based on geographic nonresponse adjustment generally are more like the unadjusted differences than either group is like the differences involving reason of noninterview. All three are relatively quite similar, however.

Summary and Recommendations

None of the discussion in this chapter attempts to suggest which type of adjustment produces the most accurate estimates or even whether the time spent doing any type of adjustment for nonresponse is time well spent. In fact in most data collection there is no practical way to find out what would be the responses of all nonrespondents, so there is no way to determine the improvement in the estimates caused by noninterview adjustment. We can only measure the change it makes in the unadjusted estimates. In our examples few of the estimates adjusted for nonresponse vary markedly from the unadjusted estimates. However, a well-thought-out plan for noninterview adjustment is usually worth making, because doing so is fairly simple. Further, benefits of noninterview adjustment are increasing because the response rates of most surveys have been declining for at least a decade (Daniel, 1975, p. 292). The larger the sample size, the more impact on the total error a noninterview adjustment will have because the standard error component of the total error will be smaller.

A fairly firm plan for noninterview adjustment should be devised before the study interviews are conducted so the desired information can be collected both for those interviewed and for those who either refused or were never found at home. The type of information collected depends upon the correlations of important variables in the study to variables that could be given values for the nonrespondents as well as for the respondents (type of dwelling, for example). The larger the correlation, the more gain noninterview adjustment produces. Categories can be chosen by using data from previous surveys to estimate correlations.

As previously stated, noninterview adjustment also requires the values to impute to the nonrespondents in each category. Unless there is firm evidence to the contrary from a fairly recent and similar survey or a pattern is found in the data according to ease of obtaining the interview, it would be best to assume that respondents and nonrespondents falling into the same category are otherwise identical. That is, such an adjustment assumes that respondents and nonrespondents differ *overall* and that this difference can in large part be explained by the differing distribution of respondents and nonrespondents between the chosen adjustment categories.

In the majority of cases it is preferable to use internal data for noninterview adjustment rather than external data. Definitional differences between the present data and the external data make it difficult to form noninterview categories. (An example of this can be seen in the fact that we are able to present data adjusted for nonresponse using external data only for demographic categories. Procedural differences usually lead to differing response rates and so on, which might result in unlike types of persons being nonrespondents in the two surveys. Thus it is unlikely that the cost and time spent locating and adapting external data could be justified.

It is difficult to choose between the two adjustment methods that use internal data, given the limited information available on the effect of each. We feel using either is somewhat preferable to performing no noninterview adjustment at all, but either set of categories could be used. When it is possible to do so, the effective nonresponse rate might be reduced by subsampling initial nonrespondents and by looking for a pattern in responses by the ease in obtaining an interview.

Because there usually is no way to determine the potential responses of nonrespondents, it is extremely difficult to decide upon the best nonresponse adjustment procedure. However, a few studies have been conducted that *do* give data estimates for both respondents and nonrespondents. (For instance, administrative records might be available for all persons designated for interview.) In these instances it can be determined if there are significant differences between survey respondents and nonrespondents and what choice of adjustment categories would be best. However, not only are such studies infrequent, but it is also rarely clear to what degree the results are generalizable to other surveys.

The amount of available information about how nonrespondents differ from respondents needs to be increased appreciably. Researchers conducting interviews with persons for whom administrative records are available should compare entries on nonrespondents' records with those of respondents. The data should also be examined by ease of getting the interview and by reason for noninterview so the results can be generalized to other surveys that may have different response rates, different patterns of reasons for noninterview, or both. The results should be made available to other researchers.

Noninterview adjustment is a relatively inexpensive method of reducing nonresponse bias somewhat, but it certainly is no substitute for obtaining actual responses from as many designated individuals as possible.

CHAPTER 8

Adjusting for Nonresponse to Specific Questions

Martin R. Frankel
Martha J. Banks

Complete noncooperation or nonavailability of designated respondents is most likely the largest source of the bias component we have designated as nonsampling bias due to nonobservation. It is not, however, the only source. Even if a respondent is willing to be interviewed, that respondent may be unable or unwilling to provide information for *every* survey question. The fact that most analytical survey software packages provide for the use of "missing data" codes attests to the prevalence of item-specific nonresponse.

Three basic methods traditionally have been used to deal with this problem: (1) Restrict variable-specific calculations to those cases for which the specific variable(s) have been provided by the respondent. (2) Do not accept any data cases that contain missing data. In other words, restrict variable-specific calculations to those cases for which *all* variables have been provided by the respondent. (3) Use a procedure to "impute" missing data values on a case-specific or subgroup-specific basis. In the health care and expenditures study described in this book, a form of this method was used to deal with item-specific nonresponse for certain important variables.

Although it may not be immediately obvious, all three methods involve a form of "data adjustment." In fact, a strong case can be made that the third method, which initially appears to involve the most adjustment, is in fact the most conservative of the available methods. If we do not use the third method, we are in fact imputing the mean (median, and so on) value among all cases for which a response exists to all cases where responses are missing. Even where the relationship between the variable in question and other survey variables used in imputation is weak, the implicit imputation induced by the elimination of case-specific nonresponse from the analyses (the first method) may be inappropriate. (This statement refers to univariate analyses. The case against elimination of such cases is not nearly as strong when we consider estimates of relationships.)

This is illustrated by the following example. Consider a population composed as follows:

	Item-Specific Respondents		Item-Specific Nonrespondents	
	Number	Mean Value	Number	Mean Value
Group 1	3	100	2	95
Group 2	4	110	1	115

The first method would result in an estimated overall mean of $[(3 \times 100) + (4 \times 110)] \div 7 = 105.7$; the true mean is $[(3 \times 100) + (4 \times 110) + (2 \times 95) + (1 \times 115)] \div 10 = 104.5$. A simple form of the third method would yield a mean of $[((3 + 2) \times 100) + ((4 + 1) \times 110)] \div 10 = 105.0$. Thus the third method produces less imputation bias in this case than does the first. However, the true mean of 104.5 is not normally known, so a case could be made that differences between variable-specific responders and nonresponders were such that the estimate using the first method (105.7) might be closer to reality than the mean obtained by group-specific imputation (105.0). Such an argument would, however, require substantial outside evidence.

Effect of Imputation on Hospital Expenditures

We have chosen to examine the effect of imputation on the variable "hospital expenditures" because this variable required a great deal of imputation. As a result we felt that differences among alternative forms of imputation would be relatively pronounced. Of all hospital admissions, 32.0 percent (weighted) required imputation. The dollar figures imputed for these admissions represented 35.1 percent of all hospital admission dollars.

In Table 8.1 we show the number of cases for which imputation of hospital expenditures was necessary. These numbers are provided for the total sample and various population subgroups. This table also shows the percentage of all admissions (weighted and unweighted) requiring imputation as well as the percentage of total expenditure dollars these groups represent.

Method Used in the Original Imputation Procedure

The original imputation method (the third) involved use of external data sources. We obtained two tapes from the American Hospital Association (AHA)—one gave hospital characteristics and the other contained cost information. A per-diem inpatient charge for each hospital was generated from these tapes. If a hospital did not appear on the AHA tapes, data from the American Hospital Association (1970) was used to determine a per diem according to the type of facility (federal, long-term care, and so on).

To impute the cost of a hospital admission when such information was not reported, we multiplied the per diem by the number of days the respondent reported being in the hospital. When the length of stay was not reported, NCHS data on average length of stay by diagnosis was used to impute days and the per-diem rate was applied. Imputation was rare for hospital days, but frequent for hospital expenditures.

Alternative Procedures

The third imputation method used "external" data in the imputation procedure, that is, data external to the social survey and subsequent validation. There exists another set of data imputation techniques based on the use of "internal" survey data. (Much of the basic development of internal imputation techniques took place at the U.S. Bureau of the Census.) In many situations the use of internal survey data is necessary (if imputation is to be done at all) because external data are not available, are of dubious quality, or involve definitions that differ widely from the internal data. It is only when reasonably good, compatible outside data are available that the choice of imputation data source is relevant. In this case it may *still* be better to use internal data. The term "good" is relative. The good outside data may be based on averages. To the extent that persons with item nonresponse are not "average," the use of internal imputation methods may be preferable.

We examined the effect on mean hospital expenditures for three related subclass-specific imputation techniques using internal data. One of the procedures used the cell mean approach; the other two were based on alternative forms of the so-called "hot deck" method. The same set of classification cells was used throughout, so differences in results are due solely to the differing methods.

Choosing Imputation Categories

Imputation categories should be chosen in a manner that maximizes the homogeneity of cases within the category with respect to the variable to be imputed and maximizes the heterogeneity of cases in different categories.

In choosing imputation categories for use with estimates of total hospital expenditures, we began with about a dozen variables we felt could be used for prediction. Looking at the stays with expenditures reported by the respondents, length of hospital stay was by far the best predictor of stay cost. We therefore decided that one of our options would be to use length of stay as the sole determinant of imputation category. As an alternative we considered length of stay combined with residence (SMSA, nonSMSA) and another variable, which for pregnancy-related stays was a three-category variable telling reason for stay (minor, moderate, serious) and for all other stays was number of operations (none, one, two or more). Data were collapsed such that each of the two possible sets of imputation categories contained twenty-six cells. In the former set of categories this simply meant a collapsing by length of stay. In the latter set it meant a more complex collapsing; the data were categorized according to fairly complicated combinations of criteria. Initial categories were combined based on trends in the mean expenditures of cases that had expenditure information reported.

We used the latter method of category designation in preference to that involving only length of stay. The categories created using other variables as well not only produced larger correlation statistics but also increased the number of reported stays relative to the number of imputed stays in the categories in which each single imputation would be expected to have the greatest impact—the categories with the largest expenditures.

Using Imputation Categories

Given the twenty-six cells, the actual imputation is itself quite straightforward. For the cell mean method we first computed the mean expenditures per admission for each cell. (These calculations were, of course, restricted to those admissions where expenditures were reported.) The resulting cell means were then used as the values assigned to all admissions in the cell where expenditure data were not available.

The hot deck technique was carried out by first ordering admissions on the basis of the interview's geographic location. Two alternative orderings were used: (1) in PSU order, that is, east to west among SMSAs followed by east to west among nonSMSAs, and (2) in reverse PSU order. For each admission where expenditure data were not available, a backward search was in effect carried out until the first non-missing-data admission falling into the same classification cell was encountered. The expenditure value associated with this non-missing-data admission was used as the imputation value. (In actual practice the entire ordered data set is passed through an algorithm that saves in storage the most current non-missing-data value associated with each cell. When a missing-data case is encountered, reference to the appropriate cell storage provides the required value.)

Comparison of Results by Imputation Method

Table 8.2 shows the overall estimates of expenditures per admission that result from the different forms of item imputation. These results are shown for the entire sample and for the standard analytical groups used throughout this book. Table 8.3 shows the impact of these methods on the standard subgroup comparisons used in the first chapters. Given the large proportion of the sample for which imputation of hospital expenditures was necessary, the differential effect of the three methods is remarkably small.

For the 74.3 percent (weighted) (65.0 percent, unweighted) of the imputed hospital admissions for which verification expenditures were available, the mean expenditures estimates were as follows:

			Internal Imputation		
		Original		Hot Deck	
		(External)	Cell	PSU	Reverse PSU
Type of Estimate	Verification	Imputation	Mean	Order	Order
Mean expenditures estimate	$677	$694	$778	$746	$727

Thus, for those cases where outside verification was available, the use of external imputation (in other words, the procedure actually used) produces overall mean expenditure estimates closer to the results obtained by verification than any of the internal imputation methods.

The general tendency for the hot deck method to produce results closer to the external imputation results than the cell mean method is moderately surprising. We had expected the opposite result.

The general reason often given for preference of the hot deck method over a cell mean approach involves the assumption that the hot deck method will better preserve "relationships" between imputed and nonimputed variables. It is often conjectured that the cell mean approach will either attenuate or artificially strengthen these relationships. The apparently superior performance of the hot deck method with respect to the imputation itself is a welcomed result.

We are still somewhat puzzled by the relatively large differences between results obtained by the two different PSU orderings used in the hot deck procedure. PSU ordered low to high represents general east to west order. The general tendency for this ordering to produce higher imputed expenditures than the west to east ordering (reverse PSU ordering) probably reflects a geographic relationship with respect to hospital charges.

On strictly empirical grounds the reverse PSU ordering seems to be the preferable choice for this data. We expect, however, that some type of averaging method actually would be preferable. Such an approach might impute a data value equal to the mean of the values of the closest within-category respondents both preceding and following the respondent for which imputation is required.

Recommendations

Our investigation of alternative imputation approaches is limited. Cost and time considerations might make the choice among imputation procedures more difficult than the preceding suggests because locating, adapting, and using external data is much more laborious than using internal data for imputation. Additionally, using these same methods of internal imputation on other variables might produce a very different assessment of the relative merit of each. We urge further investigation.

CHAPTER 9

Weighting the Data

Martha J. Banks

Basically there are two situations that may cause a researcher to weight survey data. One of these occurs when the data resulted from a sample that was not self-weighting; the weights adjust for differential selection probabilities. The question of interest in this situation is not whether weights should be applied but how the survey results compare with those that could have been expected from a self-weighting sample. In contrast to the use of such weights, post-stratification weights are optional, regardless of the sample design. They are based upon actual survey results rather than on probabilities established before the interviews were collected.

Adjusting for Unequal Selection Probabilities

Reason for Unequal Selection Probabilities

In order to increase the reliability of data from persons and families of special concern in health policy formation, the sample design used consisted of four subsamples, each with its own sampling rate: (1) The U sample consisted of households chosen in 73 segments in the NORC master sample. These segments were so designated because the 1960 U.S. census data indicated they contained a high proportion of poor urban families. (2) The A sample was selected from each of the remaining 671 segments in the NORC national probability sample. (3) The S sample consists of families obtained by screening households in the 734 segments mentioned in the two preceding points. Households were screened for poor families and for families containing one or more persons sixty-six years or older. (4) The R sample was obtained from 193 rural segments in thirty primary sampling units chosen especially for this study. No screening procedure was used for this sample.

Because of the complex sampling design, differential weighting necessarily was applied before representative estimates could be produced. Weighting was necessary to adjust for the different probabilities of selection among sample observations. Adjustment also was made for the varying completion rates among the various subsamples (the latter adjustment was mentioned in Chapter Seven). The weights were further revised to equalize (between samples) the weighting of households in the same age-income-residence category. They also adjust for the fact that many of the households originally chosen in the S sample were not interviewed because they fit neither the income nor the age criterion. The results will be referred to as the "intermediate weights" because a further adjustment (discussed later in this chapter) was made. Table 9.1 gives all these initial and intermediate weights and the unweighted number of interviewed households in each category.

Comparing the Sample with a Self-Weighting Sample

There are several possible approaches to analyzing the effect of the differential sampling and weighting upon the reliability of the data. One of these is to construct a self-weighting subsample from the data; another is to adjust the data to simulate a self-weighting sample.

There are two distinct ways to form the largest possible self-weighting subsamples. One consists of the 1,119 A-sample households plus 1,376 ÷ 13.96462 or 99 of the U-sample households. The other consists of (1.00000 × (748 + 108) ÷ 15.71088) U-type urban poor, old households, or both, (1.12505 × 628 ÷ 15.71088) other urban households, and so on. These subsamples consist of 1,218 and 1,208 households, respectively. Thus, although this approach has the intuitive advantage of creating an actual self-weighting sample rather than a construct, this advantage is negated because the size of the largest possible self-weighting subsample is only 31 percent of the entire sample. The reliability of the estimates and estimated standard errors is reduced by $\sqrt{.31}$, that is, by .56. We therefore turn to the approach that adjusts the sample data to simulate a self-weighting sample.

A self-weighting sample would be expected to contain approximately the same relative number of persons of each possible category as exist in the entire country. The weighted number of sample cases in the actual survey (given in Appendix C) is an approximation of the relative number of persons in each of the various classes of the variables of interest. Thus the first assumption made in converting the data into a self-weighting sample was that the weighted number of cases given in Appendix C reflects the relative number of sample cases that would result from a self-weighting sample. The next assumption was that a self-weighting sample about 6 percent larger than the weighted sample could have been obtained for the same cost. This is based upon an estimate of the cost of constructing the R sample, screening households in the S sample, and the additional data-management procedures necessary in the office with a weighted sample.

The result of these two assumptions appears in Table 9.2. Because persons in old and poor households and rural farm and central city residents were oversampled, a larger proportion of the unweighted number of cases from the weighted sample came from these groups than would be expected from a self-weighting sample. Nonwhites tend to be poor, central city residents, or both, so they are also overrepresented in the sample. Fewer sample persons saw a physician than would be the case in a self-weighting sample, but more had one or more hospital admissions. The weighted sample contained only 7,132 sample persons who had a physician visit; a self-weighting sample would have contained about 8,293 (16 percent more). There would be little or no difference in the number of persons with hospital admissions; but because a self-weighting sample would have fewer persons with multiple admissions, the weighted sample would contain data for about twenty-nine more admissions.

Table 9.3 contains the number of sample persons who saw a physician and the number of hospital admissions of those in the weighted sample. The table also gives the corresponding figures for an equal-cost self-weighting sample. The standard error estimates for the estimated mean physician visits and estimated mean hospital expenditure follow each sample size figure.

The standard errors for the weighted sample were obtained as described in "Standard Error" in Chapter One. They are identical to those shown in Tables 2.7 and 3.2. The estimated standard errors for the self-weighting estimates were based on the relative standard errors and the relative number of cases obtained from each of the six strata shown in Table 9.1.* Depending on these two sets of figures, the ratio of the standard error to the number of sample cases can be higher for the actual sample or for the expected self-weighting sample. Thus in many cases the estimate from the weighted sample is based on more cases but does not have the greater reliability.

The data for physician visits show that the weighted sample improved the precision of the estimates for the poor, nonwhites, and those living on farms. A self-weighting sample would have given better results for all other estimates of mean number of physician visits, including the mean for central city residents. The data for mean expenditures per admission add persons in old households, in fair or poor health, and with three diagnoses to the group of categories for which the weighted sample improved the estimates.

*Standard error estimates for a self-weighting sample were calculated as follows for each category appearing in Table 9.3: The number of sample cases, the weighted number of cases, the estimate, and the standard error of the estimate were obtained for each of the six strata. The weighted number of cases provided the relative number of cases that would have been expected from a self-weighting sample. Formulas given in Kish (1965, pp. 90, 134, 135) were used to compute the estimated standard errors of a proportionate sample with post-stratification to the six strata totals. Note that the accuracy of these standard error estimates is less than that which an actual self-weighting sample could produce.

The standard error estimates can also be used to decide whether a self-weighting sample would have provided more precise estimates of differences between means. For example, a self-weighting sample probably would have increased by 15 percent the standard error of the difference in the mean number of physician visits between whites and nonwhites; but it would have decreased the standard error of the difference between central city and suburban residents by 30 percent.

Recommendations

To summarize, the results discussed suggest that a self-weighting sample would have been more accurate than the weighted sample for the majority of estimates. However, the estimates that were improved by the weighted sample are perhaps the most important estimates. Thus it is difficult to choose between these samples on purely technical grounds. However, there are nontechnical, political considerations that increase the utility of the weighted sample. (For example, the funders of this study wanted to emphasize and show concern for the utilization patterns of the rural population, the poor, and the elderly.) The relative importance of the different types of considerations must be taken into account when deciding whether to oversample certain population groups. Personally, I probably would choose not to oversample, at least not in a way that would require screening a potential sample respondent for the presence or absence of some characteristics. The cost of the screening procedure usually is a large enough share of the total cost of data collection that screening is probably not very efficient.

Post-Stratification Weighting

Reason for Post-Stratification Adjustment

Compared to the original population, a sample chosen from it will exhibit chance differences in nearly all possible variables. Usually there are a number of characteristics that are correlated with the dependent variables of interest in the study and for which more reliable estimates exist. Thus the study data generally can be improved by applying a set of factors that adjusts the sample distribution according to the more reliable data, though in some instances the improvement may be slight. These adjustment factors are called post-stratification or ratio estimation factors. Kish (1965, pp. 90–92), Thompson and Shapiro (1973, p. 115), and the National Center for Health Statistics (1971, p. 4) discuss them further.

Data used for calculating such factors should, of course, be based upon the very same population represented in the sample. Each of the characteristics chosen to form the categories should be fairly highly correlated with statistics of interest but not so highly correlated with each other that some are superfluous.

An upper and lower bound may be used in calculating post-stratification adjustment factors. Although doing so lessens the sample's correspondence to the more reliable information in regard to these characteristics, it controls the amount of differential weighting this set of factors causes. We used an upper bound of 2 and a lower bound of 1/2. See Appendix B for a further discussion of the purpose of these bounds.

Our Approach

Estimates appearing in all previous tables were adjusted according to the family characteristics of race, residence, size, and income. Table 9.4 contains the sixteen categories used and the factors* computed by comparing CHAS data (weighted by the intermediate weights of Table 9.1) with those of the thirty-seven-thousand-household Current Population Survey (CPS) sample for a similar time period. The CPS figures can be found in United States Bureau of the Census (1972, Table 16).

Suggested Alternative Approach

Most of the CHAS data have been examined on the individual rather than the family level. Thus it would appear to be preferable to adjust the data on the individual level. In order to

*Nonresponse adjustment should be performed before post-stratification adjustment. Thus the post-stratification factors used with the data presented in Chapter Seven differ from those given in Table 9.8.

choose categories, the mean number of disability days, physician visits, and hospital days were examined for persons based on their age, sex, race, educational level, and relationship to household head. Age and race (and sex during women's childbearing years) seemed the most relevant variables, based on their ability to form categories of persons with different mean levels of health and medical care use.

Table 9.5 indicates the seventeen suggested categories based on the individual characteristics of age, sex, and race. The category boundaries were chosen (1) according to changes in mean disability days, physician visits, and hospital days and (2) so that each category involved about the same number of unweighted sample persons, involved at least one hundred unweighted sample persons, and had roughly the same importance in the final statistics.

The post-stratification adjustment factors given in Table 9.5 were calculated by comparing CHAS data for persons (weighted by the intermediate weights of Table 9.1) with the CPS figures appearing in United States Bureau of the Census (1971, Tables 2 and 3).

Comparisons of Results

Tables 9.6, 9.7, and 9.8 present the effect of the use of post-stratification adjustment on our data. We do not intend to suggest that these specific results would occur if such adjustment were used with any or all similar data.

Table 9.6 shows the effect of post-stratification adjustment on the weighted distribution of sample persons. The most pronounced effect is on the estimate of racial composition: Both adjustments reduce the estimated percentage of the sample that is nonwhite. The original post-stratification adjustment reduces the proportion living in rural areas but the alternative increases it slightly. Among the perceived and evaluated health categories, the original post-stratification produced slightly more people in the most favorable groups than did the alternative method. The use of the original post-stratification categories also resulted in an estimate that fewer people saw a physician more than once or were admitted into a hospital. Only in the case of racial composition do we know which adjustment gives better results because this is the only case presented in which a post-stratification category is used as an analytic category.

Table 9.7 examines the effect of post-stratification on both estimates of physician visits and estimates of expenditure per hospital admission. Post-stratification seems to have little effect on physician visits. The hospital expenditure figures are more affected by the adjustments, at least by the original adjustment. The original adjustment tended to increase the mean expenditure in most categories. The main effect of the alternative post-stratification adjustment was to increase the mean for nonwhites. (The original adjustment appeared to decrease this same mean.)

Table 9.8 considers post-stratification according to its effect on the reliability of differences between groups. This table does not indicate that there was any great change in the data as a result of using either of the two sets of post-stratification adjustment factors or any measurable, consistent increase in reliability (that is, declines in standard errors of differences).

Recommendations

In order to determine definitively the effect of alternative post-stratification adjustments, we would need the results of a complete census using the sample survey questionnaire. Numerous samples could be formed from the census data; post-stratification could be performed on each; and the results could be compared with the census data and with the unadjusted sample data. We would decide whether to use post-stratification (and, if so, which set of categories to use in the adjustment) by examining the distribution of the differences between the census value and the various estimates.

It is difficult to make this same decision with the limited information actually available. We have had to examine the effect of post-stratification by comparing adjusted and unadjusted estimates from a single sample. Also, we were able to look at only two different types of estimates—mean number of physician visits and mean total hospital expense per admission.

Our attempt to measure the impact of post-stratification was further confounded by the fact that noninterview adjustment usually would be performed first but in this book we have considered each separately. Had we computed estimates using combinations of noninterview adjustments and post-stratification adjustments, some combination of the two *might* have interacted in such a way that such adjustment would have had a bigger effect than either individual adjustment would have suggested.

This information about the effect of post-stratification, limited though it is, is a useful first step in developing a more thorough investigation of the expected effect of post-stratification adjustment of different types of data.

Although it is difficult to assess the effect of post-stratification on the data and even more difficult to predict what its use would mean to other surveys, I suggest that it be done. Post-stratification is an extremely inexpensive procedure and should result in at least some small improvement in the data as long as there is more homogeneity of persons within than between categories. For this particular study, this means that persons in the same age-sex-race group or the same race-residence-family size-income group should have more similar medical experiences than do people in general. The choice of categories depends upon the nature of the survey because the categories should be delineated by characteristics correlated with important estimates in the study.

CHAPTER 10

Models for the Use of Verification Information

Martin R. Frankel

✲ ✲ ✲ ✲ ✲ ✲ ✲ ✲ ✲ ✲ ✲ ✲ ✲ ✲ ✲ ✲ ✲

Most texts that deal with sample survey design emphasize that developing "optimal" survey design consists of specifying data collection, sampling, and associated procedures that will produce the minimum total survey error for the available budget. Later chapters of those texts, however, state that optimization formulas implicitly assume the only source of survey error to be variable sampling error. The reason for this implicit exclusion is readily understandable: Other sources of variable error as well as bias are typically impossible to specify. If these other error sources are included in the development of "optimal" design, they are typically included in an ad hoc and qualitative fashion.

In order to develop survey designs that minimize the total survey error, rather than sampling error alone, we need two things: (1) mathematical models that relate the magnitude of total survey error to certain population parameters and costs and (2) reasonably good estimates of these required parameters and costs.

This chapter describes several models that may be used to develop survey designs that minimize two of the basic components of the total survey error—sampling error and response bias. We have purposely limited our models to these two total survey error components because we feel the data collection from this survey will support the necessary parameter estimates required to apply the model. As empirical studies that provide appropriate parameter estimates for other error sources become available, the models described may be suitably extended.

Basic Assumptions

In our models, we assume two available methods of data collection: a social survey questionnaire and a secondary record check procedure (verification). In order to use the verification procedure, we must first obtain data via the social survey questionnaire. We can, however, use the social survey questionnaire and not conduct a subsequent record check. In our models we assume that increasing the sample size for the social survey questionnaire alone will decrease the variable sampling component of the total survey error but will have no impact on the bias components of the total survey error. Increasing the size of the subsample for which verification information is obtained, however, may decrease both variable sampling and nonsampling error, as well as bias. Due to the lack of appropriate data inputs, the model does not attempt to measure or control the nonresponse component of the total bias.

For expositional simplicity we have assumed simple random sampling of elements. Design features such as stratification and clustering may be added to the model in a fairly straightforward manner.

Our basic model assumes that a survey instrument (questionnaire) will be administered to a sample of n population elements. Some proportion (k) of these sample elements may be subjected to verification from outside sources. Thus the size of the verification sample is kn. The symbol y_{ssi} is used to denote the value of variable Y obtained from the ith sample respondent via the social survey questionnaire. If verification information regarding variable Y is obtained for the ith sample element, its value will be denoted by y_{vi}.

It is assumed that data collection costs are the same for all sample elements. The symbol C_1 is used to denote the "per-element" cost of obtaining social survey information (y_{ssi}) for a single respondent. The additional cost of obtaining verification information for a single respondent is denoted by C_2. The total cost of data collection is $T = nC_1 + knC_2$.

On a per-element basis, the total survey error associated with a social survey response y_{ssi} is denoted by

$TSE\ (y_{ssi}) = S_{ss}^2 + B^2$, where S_{ss}^2 is the per-element variance of y_{ssi} and B^2 is the square of the response bias.

The per-element total survey error associated with a single verification observation y_{vi} is denoted by

$TSE\ (y_{vi}) = S_v^2$, where S_v^2 is the per-element variance of y_{vi}

It should be recognized that S_{ss}^2 and S_v^2 may differ from each other because y_{ssi} is also subject to variable response error. In many situations we would expect this difference to be small. To summarize,

y_{ssi} = value obtained for i^{th} sample respondent from social survey
y_{vi} = value obtained for i^{th} sample respondent from verification

C_1 = cost of obtaining y_{ssi} for a single case
$C_1 + C_2$ = cost of obtaining both y_{ssi} and y_{vi} for a single case

n = the sample size for the social survey
kn = the size of the subsample for which verification information is obtained ($0 \leq k \leq 1$)

$$\text{Total cost of data collection } T = nC_1 + knC_2$$
$$TSE\ (y_{ssi}) = S_{ss}^2 + B^2$$
$$TSE\ (y_{vi}) = S_v^2$$

where TSE denotes total survey error, S_{ss}^2 and S_v^2 denote variable errors, and B^2 denotes the square of the social survey response bias.

Given these definitions, the question to ask of the models is how to allocate the total budget T. Should the entire budget be devoted to collecting social survey information only or should the sample size be decreased to allow for collection of some verification information? The answer depends upon the relative costs (C_1 and C_2), the variable errors and bias (S^2_{ss}, S^2_v, and B^2), and the procedure for utilizing the social survey and verification data.

Substitution Model

Andersen (1975) and Andersen, Lion, and Anderson (1976) have used data obtained through verification on a case-by-case basis. That is, if both y_{ssi} and y_{vi} (social survey and verification) are available for a respondent, the latter is used to develop overall estimates.

Thus the overall survey estimate of $\overline{Y}$ (the population mean) is

$$\overline{y}_s = \frac{1}{n} \sum z_i, \tag{1}$$

$$\text{where } z_i = y_{vi} \text{ if available}$$
$$= y_{ssi} \text{ otherwise}$$

Under the assumption of simple random sampling and subsampling, the total survey error of the estimate $\overline{y}_s$ is *

$$TSE\ (\overline{y}_s) = \frac{(1-k)\ S^2_{ss} + kS^2_v}{n} + (1-k)\ B^2 \tag{2}$$

Assuming $S^2_{ss} = S^2_v = S^2$, this becomes

$$TSE\ (\overline{y}_s) = \frac{S^2}{n} + (1-k)\ B^2 \tag{3}$$

For a fixed budget T, the total survey error given by (3) is minimized when k, the proportion of cases to be verified, is

$$k = 1 - \tfrac{1}{2} \frac{C_2\ S^2}{T\ B^2} \tag{4}$$

As this formula indicates, as B, the magnitude of the response bias, increases, the proportion of cases to be verified increases. As C_2, the cost of this validation, increases, however, the proportion k goes down.

*These formulas, and those that follow, ignore the effect of the finite population correction factor.

Ratio Adjustment Model

This method, known in the survey sampling literature as two-phase ratio adjustment, makes use of the subsample of kn cases to produce an estimate of the ratio of the mean verification estimate to the mean social survey estimate. This ratio is then applied to the estimate of the social survey mean, obtained from all n sample observations, to produce the final estimate. This method differs from the substitution approach in that information obtained from the subsample is used as a correction factor for the entire sample, not on a case-by-case basis.

Let $\bar{y}_{ss}$ denote the overall social survey mean obtained from all n sample observations. Let $\bar{y}'_{ss}$ and $\bar{y}'_v$ denote the social survey and verification means obtained from the kn subsample cases only. Further, let r denote the ratio of these subsample means, $r = \bar{y}'_v / \bar{y}'_{ss}$.

The ratio adjustment estimate of the population mean is given by

$$\bar{y}_r = r\,\bar{y}_{ss} = \frac{\bar{y}'_v}{\bar{y}'_{ss}} \cdot \bar{y}_{ss} \tag{5}$$

By assuming that kn is of sufficient size so that the potential statistical bias in r may be ignored, the total survey error of $\bar{y}_r$ is given by

$$TSE\ (\bar{y}_r) = VAR\ (\bar{y}_r) = \frac{S_r^2}{kn} + \frac{(S_v^2 - S_r^2)}{n} \tag{6}$$

when

$$S_r^2 = R^3 S_{ss}^2 + S_v^2 - 2R\, S_{ss,v} \tag{7}$$

$S_{ss,v}$ is the element covariance between the social survey (y_{ssi}) and the verification (y_{vi}), and R is the population value of the ratio r. (In practice r is used as an estimate for R.)

If the per-unit variances S_{ss}^2 and S_v^2 are equal, the expression for S_r^2 becomes

$$S_r^2 = S^2\,(1 + R^2 - 2R\rho_{ss,v}) \tag{8}$$

where $S^2 = S_{ss}^2 = S_v^2$, and $\rho_{ss,v}$ is the correlation (per element) between the social survey and the verification

The optimal value of k, the proportion to be verified, occurs when

$$k = \frac{S_r}{\sqrt{S_v^2 - S_r^2}} \sqrt{\frac{C_1}{C_2}} \tag{9}$$

As this formula indicates, as C_2, the cost of verification, increases, the proportion to be verified declines. As the standard error of the correction factor r increases, however, the size of the second-phase verification in subsample is increased.

Regression Adjustment Model

The regression adjustment method is a modification of the ratio adjustment technique. Rather than adjusting the overall social survey mean by a simple ratio, the adjustment allows for an additional constant term. The first step in developing this adjustment is to specify a regression relationship between social survey responses and verification information as follows:

$$y_{vi} = A + By_{ssi} + e_i \tag{10}$$

where y_{vi} = verification value for i^{th} sample individual

y_{ssi} = social survey value for i^{th} sample individual

A, B = unknown constants

e_i = error term with expectation zero

Usual least squares procedures are applied to the subsample of individuals for which social survey and verification information has been obtained to produce estimates of the regression coefficients A and B. These estimates, denoted a and b respectively, are then applied to $\bar{y}_{ss}$, the overall social survey mean, to produce our regression adjustment estimate

$$\bar{y}_{rg} = a + b\,\bar{y}_{ss} \tag{11}$$

Letting $\bar{y}'_{ss}$ and $\bar{y}'_v$ denote the social survey and verification means developed from the subsample and noting that in the least squares estimation procedure $a = \bar{y}'_v - b\bar{y}'_{ss}$, we note that $\bar{y}_{rg}$ may be written as

$$\bar{y}_{rg} = \bar{y}'_v + b\,(\bar{y}_{ss} - \bar{y}'_{ss}) \tag{12}$$

Expression (12) follows the format used in most sampling texts.

The total survey error of this estimate is approximated by

$$TSE\ (\bar{y}_{rg}) = \frac{S_v^2\ (1 - \rho_{v,ss}^2)}{kn} + \frac{S_v^2\ \rho_{v,ss}^2}{n} \tag{13}$$

For a given budget, this total survey error is minimized when k, the proportion of the sample designated for validation, is

$$k = \sqrt{\frac{C_1\ (1-\rho_{v,ss}^2)}{C_2 \cdot \rho_{v,ss}^2}} \tag{14}$$

From formula (14) we note that as C_2, verification cost, increases, the optimal proportion of cases to receive verification declines. As $\rho_{v,ss}$, the correlation between the information obtained by the social survey and that obtained by verification, gets larger, the optimal subsample proportion k declines. In this case a relatively few number of verification cases will provide adequate data for the estimation of bias correction estimates a and b. Note that in all the optimization formulas the value of k may be outside the permissible range of 0 to 1. In such cases k is empirically constrained to be either 0 or 1.

Before providing an example of the relative efficiency of the three alternative methods, we note that the regression adjustment estimate $\bar{y}_{rg}$ will always have total survey error less than or equal to the total survey error of the ratio adjustment $\bar{y}_r$.

The ratio of total survey error for the ratio and regression adjustment methods is given by

$$\frac{TSE\ (\bar{y}_r)}{TSE\ (\bar{y}_{rg})} = 1 + \frac{\left(\frac{S_v^2}{S_{ss}^2} - \rho_{v,ss}\right)^2}{(1 - \rho_{v,ss}^2)} \tag{15}$$

This ratio of total survey error becomes one when the second term in expression (15) vanishes. In this case the regression slope goes through the origin (constant $A = 0$) and both estimates become identical.

Even though the ratio adjustment method has total survey error that is in most cases larger than that of the regression adjustment method, its use may still be appropriate and desirable. In many situations the difference in total survey error between y_r and

$\bar{y}_{rg}$ may be quite small. The ratio adjustment requires less complex calculation, and to the unsophisticated survey user its justification may seem more straightforward and reasonable.

Table 10.1 shows the total survey error values that would be obtained for the three alternative estimates ($\bar{y}_s$—case-by-case substitution, $\bar{y}_r$—two-phase ratio adjustment, and $\bar{y}_{rg}$—two-phase regression adjustment) under various sampling plans. The following situation is assumed: The total data collection budget is T = \$10,000. The cost of collecting a single "social survey" response is C_1 = \$10. The additional cost of obtaining a single "validation" response is C_2 = \$25. The variable to be estimated has true mean $\bar{Y}$ = 490. The expected value of the "social survey" responses for the population is 500, and thus bias $B = -10$. Finally, the per-unit variance of both the social survey and validation observations is assumed to be $S^2 = S^2_{ss} = S^2_v = 25{,}000$, with correlation between social survey and validated responses $\rho_{ss,v} = 0.9$.

Each row of Table 10.1 shows a sampling allocation of social survey (total sample) and validated (subsample) responses with total cost \$10,000. For each allocation total survey error values are shown. For example, in the fifth row of the table, the sampling allocation plan calls for a sample size of 600 cases with social survey responses and a subsample of 160 "validated" responses. The total cost of the plan is (600 × \$10) + (160 × \$25) = \$10,000. If we used the mean substitution estimate, the total survey error would be TSE ($\bar{y}$) = 115.00. The two-phase ratio adjustment estimate would have TSE ($\bar{y}_r$) = 65.10, and the two-phase regression adjustment would have TSE ($\bar{y}_{rg}$) = 63.44.

In the first row of the table the total survey error is shown only for the substitution estimate $\bar{y}_s$. In this case there is no subsample and thus the ratio or regression adjustments are not possible. The substitution estimate is simply the mean of the social survey responses.

For the parameters assumed, the two-phase regression adjustment estimate has the lowest attainable mean square error, TSE ($\bar{y}_{rg}$) = 63.44, when k = .26667. The ratio adjustment estimate has the next lowest attainable mean squared error, TSE ($\bar{y}_r$) = 65.10, when k = .26667. Both regression and ratio adjusted estimates are clearly superior to the substitution estimate, which has the minimum TSE ($\bar{y}_s$) = 87.72, when k = 1. Note that the optimal mixture of social survey and validated cases is quite sensitive to changes in the ratio of costs C_1/C_2. In the actual health care survey this ratio was quite large (in other words, the additional cost of validation was small relative to the cost of the social survey). In such situations the optimal proportion of cases for which validation should be sought will be equal to unity under a wide variety of conditions for the other parameters.

Table 10.2 estimates social survey and verification means, variances (per element), and covariances for two of the most important variables studied in the Health Access Survey: hospital expenditures and physician visits. Note that these values do not include any imputed values.

The differences in the correlation between social survey and validation responses for the two variables (.94 for hospital expenditures versus .62 for physician visits) illustrates one of the complexities that must be faced when developing an overall design strategy. Under the assumption that the regression adjustment procedure will be used, formula (14) indicates that the optimal subsampling fraction k is equal to C_1/C_2 (.3787) for hospital expenditures and C_1/C_2 (1.2816) for physician visits.

In order to resolve differences in optimal design parameters, the researcher must assign relative weights to the importance of different variables and simultaneously examine the anticipated total survey error of the various estimates throughout the range of possible subsample values. In a reasonably broad neighborhood around the optimal point, total survey errors often will differ only slightly. This is similar to the situation that exists in the development of optimal allocation for stratified sampling.

CHAPTER 11

Methodological and Social Policy Issues

Ronald Andersen

This book addresses questions concerning the reliability and validity of population characteristics estimates made from large-scale social surveys. Issues considered include (1) those of making estimates of population characteristics from social surveys that consider bias as well as random error, (2) those of general concern in doing empirical analysis for social policy, and (3) those of particular relevance in health services research. This chapter reviews findings and discusses implications concerning each of these topics. The discussion of bias and random error draws from all previous chapters; the implications for social policy and health services research are based primarily on the findings in Chapters Two through Six.

Methodological Issues

The main methodological effort was to provide an example of how the concept of total survey error could be given an operational, quantifiable definition using data from a national survey. The validation studies of the medical care survey respondents reported allowed us to calculate certain kinds of bias in addition to providing estimates of random error traditionally provided in careful survey research. These biases were then used to adjust estimates of population characteristics. Biases calculated were due to nonresponse errors, characteristics in respondent reporting, and processing errors resulting from imputation of missing data. The assumption behind this undertaking was that bias may well contribute more to total survey error than does random error for some population estimates.

Although this effort has been preliminary and exploratory, we feel it is justified because of the current lack of empirical studies on the topic. This volume suggests how processes can be developed for estimating some biases in large-scale social survey research. The bias adjustments did not materially affect the estimates for many characteristics (such as total expenditures for inpatient admissions). In other instances the adjustment was considerable (such as emergency and outpatient department charges). Survey methodologists should seriously consider validating key estimates, given the possibility of substantial bias. This process can be very expensive, but Frankel's suggestions in Chapter Ten on how validation can be done on a sample basis could cut these costs considerably. It is also conceivable that as information on the magnitude and stability of biases accumulates, it will be possible to adjust estimates from a given social survey for bias using external information without needing to conduct a special validation study for that particular survey.

Our experiences suggest a number of potential improvements in the way the total survey error model is made operational and in adjusting the survey estimates. A basic assumption in our adjustment process was that the various error components were additive. In fact, some kinds of error may be interactive. For example, errors in processing the data could be influenced by the quality of the initial interview as measured by the reporting error. It would seem useful to consider explicitly if, and when, an interactive model is necessary.

Further, the error measures used in this study, although conceptually important, represent only a part of the reporting and processing components that should be considered. For example, no effort was made to measure the impact of editing, coding, keypunching, or cleaning on the processing error.

A major problem in this study was the inability always to identify false negatives (persons who did not report all their medical care experiences). This is of particular concern with contact measures such as percentage seeing a physician in a given year or number of hospital admissions per one hundred persons per year. Identifying false negatives would seem to be less important for estimating the intensity of action as represented by average length of hospital stay or dollar expenditures for a hospital admission. The bias from false negatives can be studied by placing greater emphasis on record studies where the fact of service (such as hospital admission) is established a priori and is then compared with the respondent report. The results of this type of study might be used to adjust systematically for underreporting in social surveys.

A dilemma in calculating biases is what to do about nonrespondents. Banks and Frankel point out in Chapter Seven that to ignore them is not a conservative approach because the implicit assumption is that their behavior is exactly like that of the respondents. This study sought information from external sources on the medical care use of nonrespondents and used this information to estimate the bias in our own study. As pointed out in Chapters Two and Three, the time, sample, place, and questions of the external sources do not match too well with those of our study, which makes our bias adjustments for nonrespondents somewhat suspect.

An alternative approach suggested by Banks and Frankel (Chapter Seven) is to use the study's own data on respondents for adjustment purposes. Basically, this approach assumes that the nonrespondent is like his or her responding neighbor or that nonrespondents are like reluctant respondents on characteristics we wish to estimate.

Use of internal data does away with the often thankless task of hunting for information on nonrespondents, which even if found may be only grossly related to estimating needs. However, the internal data approach also gives cause for concern if there is reason to believe that nonrespondents may differ considerably from respondents on characteristics to be estimated. Suppose, for example, that refusals in health care surveys either have so much illness and medical care that it is painful to talk about or so little need for medical care that the topic is not salient. In either case nonresponse adjustments using internal data could be misleading.

Again, more emphasis on the type of record studies mentioned earlier could be a way out of this dilemma. If a sample is drawn from lists of patients with known medical care experiences, inferences might then be drawn about the characteristics of the nonrespondents in a social survey. Issues of patient confidentiality might hinder this approach. However, even if it were not possible to link the specific medical records to individual nonrespondents, it might still be possible, without violating confidentiality, to describe the social and health care characteristics of the nonrespondents at the aggregate level. Given the dearth of available information, almost any efforts in this area would be an advancement. Further, as more information on the differences between respondents and nonrespondents became available, more informed decisions could be made about the advisability of using internal data from one's own study to estimate nonresponse bias.

Empirical Analyses in Social Policy

What this book says about general social policy analysis stems from one basic assumption: Judgments about equity in the distribution of services and the need for programmatic changes are usually based on comparisons of population subgroups. With this assumption in mind, much of the analysis in this book has been devoted to comparisons according to age, income, race, residence, and health status. How do presumed "disadvantaged" groups, including the elderly, poor, farm and central city residents, and those in poor health, compare with others in the population? Are these groups of particular policy interest receiving more or fewer services than those thought to be "better off"?

Our concern was that biases might be correlated with these characteristics so that apparent differences among subgroups based on social surveys could be misleading. It was conceivable that actual differences could either be exaggerated or understated because of the presence of correlated bias.

We found rather few instances where the bias adjustments change the conclusions that would be drawn about the statistical significance of differences of health care experiences between subgroups. However, the magnitudes of the differences do change more often. Such changes can lead to altered inferences about the relative status of subgroups. (A "change" in the difference is assumed to be of significance for policy purposes if it exceeds the standard error of the difference between groups being compared.)

Table 11.1 summarizes the impact of bias adjustments on selected indicators of inpatient care, ambulatory care, and health insurance coverage. The resulting picture is mixed. In some instances the bias adjustments appear to change the relative status of the elderly, the poor, minorities, farm and central city residents, and those in poor health. In other cases the bias adjustment has little impact on the standing of these groups of policy interest relative to the comparison groups noted in Table 11.1.

The inpatient care comparisons are, overall, least affected by the bias adjustments for all the groups of policy interest. The picture is not appreciably altered for hospital days, hospital expenditures, or out-of-pocket costs for hospital care—with one exception. The bias adjustment suggests the elderly pay relatively more out-of-pocket than the unadjusted comparisons would indicate.

The one characteristic of the hospital experience that is substantially altered for all the policy groups except the farm population is the reporting of conditions for which patients were hospitalized. The elderly, poor, nonwhites, central city residents, and those in poor health all appear to be relatively inaccurate reporters. Consequently, the validation of diagnoses tends to improve reporting accuracy for these groups more than is the case for their comparison groups.

The bias adjustments generally have more impact on comparisons for the ambulatory care indicators than the hospital indicators. As a result of the bias adjustments all the policy groups except central city and farm residents appear to make fewer visits relative to other population groups than the unadjusted comparisons indicate. These results suggest that the policy relevant groups may not be "as well off" as the social survey alone would indicate, assuming relatively more visits for these groups is the desired state.

Expenditures for visits to physicians' private offices are less affected by the bias adjustments than are number of visits. However, for central city residents and those in poor health the bias adjustments reduce expenditures relative to the comparison groups.

The effect of the bias adjustment for outpatient department and emergency room expenditures is the opposite of what it was for expenditures in private doctors' offices. For all policy groups except the elderly the bias adjustment *increases* the expenditures relative to the comparison groups. It appears that reliance on social surveys alone would seriously underestimate the expenditures for outpatient and emergency room services received by minority groups.

As was the case with hospital conditions, the accuracy of reporting of conditions for which a doctor was seen generally increases for the policy groups more than for the other groups as a result of validation. The exceptions to this generalization are the aged and those in poor health whose original reporting accuracy was as good as that of the comparison groups.

The impact of the bias adjustments for health insurance measures was more varied than for hospital and ambulatory care measures. The bias adjustments did not affect the comparison of premiums families paid for health insurance except that the elderly apparently pay relatively more than the social survey indicated.

The adjustment process generally resulted in the "disadvantaged" groups having relatively more coverage provided by their insurance policies for hospital care and physician visits. The nonwhites were an exception in that the adjustment made no difference for hospital coverage and actually reduced their physician visit coverage relative to the whites.

All groups in the population had trouble identifying major medical coverage in their policies. Apparently, the "disadvantaged" groups (again with the exception of the nonwhites) had more problems and they tended to report major medical coverage that was not verified. Consequently, the bias adjustment reduced their major medical coverage relative to the comparison groups.

In sum it appears that some biases are indeed correlated with social characteristics of major importance in social policy analyses. Although in relatively few instances do bias adjustments lead to different conclusions about whether differences between subgroups are statistically significant, they more often result in altered perceptions about the degree or extent of differences between the groups of policy interest and their comparison groups. Further, with a number of exceptions previously noted, the various subgroups considered here (the elderly, poor, nonwhite, farm and central city residents, and those in poor health) are affected similarly by bias adjustments. That is, when bias adjustments for indexes of inpatient care, ambulatory care, or health insurance change the comparative status of one of these groups, similar changes often occur for the other groups as well. In doing comparative analyses of subgroups for policy research, such potential biases should be considered.

Health Services Research

Much of the previous discussion of methodology and social policy analysis issues is also relevant to health services research. However, this section will not concentrate on subgroup comparisons but will review some general conclusions concerning health survey data collection. A major focus will be whether bias adjustments tend to increase or decrease the gross estimates of incidence or prevalence of health service experiences. Also of concern is the extent to which there is apparent inaccuracy in the social survey estimates that is not reflected in the gross bias adjustments. That is, there could be major inaccuracies at the individual observation level, with many survey estimates being too high and others too low. However, the net bias might be negligible, with the positive and negative biases canceling each other.

Two qualifications made throughout this book need to be reiterated. One is that the design is not efficient in uncovering false negatives (persons who had medical care and did not report it). Consequently, we will focus on volume estimates, which are less likely to be affected by the lack of false negative information. Thus, findings discussed will include days per hospital admission rather than admission per one hundred persons per year, number of physician visits for persons with visits rather than percentage of persons seeing a doctor within a year, and benefits provided by a given insurance policy rather than percentage of the population with insurance.

A second qualification is that we are generally accepting the reporting of hospitals, physicians, and insurers as the validity criteria. To the extent those reports are incomplete or inaccurate, our conclusions can be misleading. However, although the precise extent of inaccuracy in the third-party reports is not known, we judge them to be acceptable for the purposes of this study. This judgment takes into account the current state of the art characterized by limited empirical evidence and our efforts to control the quality of validation information during our collection and processing activities.

Table 11.2 reports the proportional bias and the extent of underestimating and overestimating in the social survey for the same estimates of inpatient care, ambulatory care, and health insurance considered in the last section.* The proportional bias shows the social survey estimate as a proportion of the adjusted estimate. The proportion underestimated or overestimated shows the proportion of observations used to make the adjusted estimates that have values in the social survey less than (underestimated) or more than (overestimated) the values in the adjusted estimates. For inpatient care the unit of observation is the hospital admission; for ambulatory care it is the patient receiving services; for health insurance it is the insurance policy.

The social survey estimates exceed the adjusted estimates for two of four selected indexes of inpatient care and are less than the adjusted estimates for the other two. The social survey estimates are higher for average length of stay and are much higher for out-of-pocket expenditures per admission. In both cases it appears that individuals are considerably more likely to exaggerate than to underreport their length of stay and what they must pay for inpatient care.

However, the social survey estimates of total expenditures

*Underreporting and overreporting were examined in Chapter Six using the person as the unit of analysis. Only reported data were used in that analysis, but Table 11.2 shows both reported and imputed data in comparing underestimating and overestimating.

for a hospital admission are slightly *less* than the adjusted estimates because people seem to underreport third-party (voluntary insurance, Medicare, Medicaid) payments for hospital care but overreport their own contributions. Further, Table 3.2 showed that although there was no overall field bias in the total expenditure estimate, there was a considerable nonresponse bias (based on the assumption that nonresponders had more expensive hospitalizations than responders). It is the correction for this nonresponse bias that makes the adjusted estimate higher than the social survey estimate. Table 11.2 also indicates that persons underreport the conditions responsible for the hospitalization as compared to diagnoses given on hospital records.

Three of the four ambulatory care indexes are overestimated in the social survey (Table 11.2). Given that they have seen physicians, respondents report more visits than can be verified through physicians' records. The survey estimates of expenditures for visits to doctors' private offices are also somewhat higher (8 percent) than the adjusted estimates. A review of Table 3.4 shows that doctor visit expenditure estimates are higher in the social survey due to both excess respondent reporting and to higher imputation values for "no answers" than found in the verification.

Respondents also tended to report more conditions for which they saw a doctor than were obtained in the record checks. Table 5.2 shows that the overreporting is largely composed of nonspecific reasons for seeing a physician such as "checkup" and "tests and shots." Most reasons involving a diagnosis were actually reported more frequently in the verification than by social survey respondents.

The one ambulatory care measure for which the social survey estimate is considerably less than the adjusted estimate is expenditures for outpatient department and emergency room services. The findings, however, do not indicate that survey respondents were underreporting expenditures for outpatient services. Rather, in many instances they were unable to give any information on the cost of this care. In such instances our office staff imputed values for the social survey estimates. Apparently, our imputation procedures for emergency room and outpatient department care were conservative, resulting in underestimates of the value of these services as determined by the verification studies.

All selected health insurance characteristics—family premium costs for the policy and whether the policy covered hospital costs, doctor visit costs, and major medical expenses—were overestimated in the social survey. Respondents appeared particularly likely to overreport family premiums (although a significant minority also underreported premiums) and major medical coverage. Among all the types of coverage studied, only dental coverage seemed to be underreported in the social survey (Table 4.9).

In sum, the assumption that respondents in health surveys generally underreport, particularly with recall periods as long as a year, can be called into question by the results of this study. The opposite seemed to be true for many of the measures we examined. Further, in some instances a social survey underestimate resulted not from respondents underreporting but from other types of biases, such as nonresponse bias in the case of expenditures per admission and processing (imputation) bias in the case of outpatient and emergency room expenditures. One possible explanation for respondents not necessarily underreporting when faced with long recall periods, with the need to report on events that require detailed information, and with happenings that may not be highly salient concerns how they arrive at their answer. Possibly rather than attempting to recall each unique event, for example, each physician visit made over the past year, a respondent may use an estimating technique based on "usual behavior" concerning preventive care and response to episodes of illness. Such an estimating technique could conceivably result in *either* an underreporting or overreporting of health care experiences. In any case it would seem judicious given the current state of the art to be sensitive to the possibility that bias corrections of social survey estimates may go "either way."

Tables

Table 1.1: Estimated Response Rates by Demographic Characteristics

Characteristic	*Estimated Response Rates* Respondents $(1-p)$	Nonrespondents (p)	Total
Age of oldest family member (7)[a]			
Less than 65 years	81.7%	18.3%	100.0%
65 years or more	84.5	15.5	100.0
Family income (26)			
Nonpoor	81.6	18.4	100.0
Poor	83.0	17.0	100.0
Race (62)			
White	82.0	18.0	100.0
Nonwhite	81.5	18.5	100.0
Residence (64)			
Rural nonfarm	84.4	15.6	100.0
Rural farm	89.7	10.3	100.0
SMSA central city	78.1	21.9	100.0
SMSA other urban	81.6	18.4	100.0
Urban nonSMSA	82.9	17.1	100.0
Total	82.0%	18.0%	100.0%

[a]In this table and all subsequent tables in the chapter, the numbers in parentheses refer to the variable definition list found in Appendix C.

Table 1.2. Verification of Hospital Admissions, Physicians Seen, and Health Insurance Policies

Unit of Observation	*Reported in Social Survey*					*Discovered in Verification*
	Total [1][a]					[6]
		Verification Attempted				
		No [2]	*Yes*			
			Verification Completed			
			No [3]	*Yes*		
				Observation Accepted		
				No [4]	*Yes* [5]	
Hospital admissions	1,771	143	70	163	1,395	110
Physicians seen	10,981	1,678	1,567	1,639	6,097	168
Health insurance policies in effect	3,635	356	298	181	2,800	65

Note: All data are unweighted.

[a]Column 1 = Column 2 + Column 3 + Column 4 + Column 5.

Table 2.1. Comparison of Social Survey Estimates and Verification of Hospital Days per Admission

Type of Comparison	*Percent of All Admissions*			
	Social Survey Equals[a] Verification	*Social Survey Exceeds Verification*	*Verification Exceeds Social Survey*	*All Admissions*
Difference in days				
0	39.9%	—%	—%	39.9%
1 to 2	—	21.0	7.8	28.7
3 to 5	—	9.7	3.6	13.3
6 to 10	—	8.5	4.7	13.2
11 or more	—	2.9	1.9	4.8
All differences	39.9	42.1	18.0	100.0
Percent difference				
1.0% or less	40.0	—	—	40.0
1.1 to 5.0	—	1.3	.8	2.1
5.1 to 25.0	—	12.0	6.0	18.0
25.1 to 75.0	—	15.3	5.9	21.2
75.1 or more	—	6.5	.3	6.8
Rejected in verification	—	7.0	—	7.0
Discovered in verification	—	—	4.9	4.9
All differences	40.0	42.1	17.9	100.0

Note: Admissions for which we received no response from any verification source were excluded.

[a]*Equals* means within one day for the first category in this column and within one percentage point for the second.

Table 2.2. Error Components in Social Survey Estimates of Hospital Admissions per 100 Person-Years

Characteristic	Hospital Admissions per 100 Person-Years (30)				
	Social Survey Estimate	*Standard Error*	*Nonresponse Bias*[a]	*Field Bias*[a]	*Processing Bias*[a]
Demographic					
Age of oldest family member (7)[b]					
Less than 65 years	13.3	.7	−.1	− .3	—[c]
65 years or more	20.7	1.6	−.1	−1.5	—
Family income (26)					
Nonpoor	13.8	.8	−.1	− .3	—
Poor	16.2	1.0	−.1	− .8	—
Race (62)					
White	14.7	.7	−.1	− .4	—
Nonwhite	11.5	.9	−.1	− .8	—
Residence (64)					
Rural nonfarm	15.0	1.2	−.1	− .2	—
Rural farm	14.5	1.2	−.1	−1.0	—
SMSA central city	12.5	1.0	−.1	− .7	—
SMSA other urban	14.7	1.7	−.1	− .3	—
Urban nonSMSA	16.3	1.2	−.1	− .7	—
Perceived and evaluated health					
Perception of health (55)					
Excellent	7.2	.7	—[d]	− .1	—
Good	13.7	.9	—	− .2	—
Fair	22.9	2.2	—	−1.6	—
Poor	47.1	4.4	—	−3.3	—
Worry about health (73)					
A great deal	59.0	4.1	—	−2.8	—
Some	22.6	1.6	—	− .6	—
Hardly any	9.7	1.0	—	− .4	—
None	5.6	.5	—	e	—
Severity over all diagnoses (67)					
Elective only	3.1	.4	—	e	—
Elective and mandatory	31.3	2.1	—	−1.6	—
Mandatory only	35.1	2.1	—	− .4	—

Table 2.2. (Continued)

Characteristic	Hospital Admissions per 100 Person-Years (30)				
	Social Survey Estimate	*Standard Error*	*Non-response Bias*[a]	*Field Bias*[a]	*Process-ing Bias*[a]
Number of diagnoses (45)					
1	12.8	1.2	—	− .3	—
2	18.7	1.3	—	−1.0	—
3	28.6	2.5	—	−1.7	—
4 or more	52.5	4.3	—	−1.2	—
Total	14.3	.7	−.1	− .4	—

Note: An earlier version of some of the above data appeared in Andersen (1975). Differences between the two tables are due to the inclusion of long-term hospital stays in this table and their exclusion from the earlier table.

[a]In this and all subsequent tables in the chapter, a positive bias figure indicates that the amount should be added to the social survey estimate to reach the adjusted estimate, and a negative figure indicates that the amount should be subtracted.

[b]In this and all subsequent tables in the chapter, the numbers in parentheses refer to the variable definition list found in Appendix C.

[c]Fact of hospital admission was not imputed.

[d]Nonresponse biases were not calculated for perceived and evaluated health variables because necessary external data were not available.

[e]Less than .05 admissions.

Table 2.3. Error Components in Social Survey Estimates of Hospital Days per Admission

	Hospital Days per Admission (33)				
Characteristic	*Social Survey Estimate*	*Standard Error*	*Non-response Bias*	*Field Bias*	*Process-ing Bias*
Demographic					
Age of oldest family member (7)					
Less than 65 years	8.9	.7	.3	− .5	a
65 years or more	16.0	1.6	.5	− .6	0
Family income (26)					
Nonpoor	9.8	.8	.3	− .6	a
Poor	11.9	1.2	.4	− .4	a
Race (62)					
White	10.3	.7	.4	− .6	a
Nonwhite	10.4	.9	.4	− .3	0
Residence (64)					
Rural nonfarm	10.2	1.4	.3	−1.2	a
Rural farm	10.1	1.2	.2	.2	0
SMSA central city	10.4	.9	.4	− .3	0
SMSA other urban	12.0	2.1	.4	− .7	a
Urban nonSMSA	7.4	1.1	.2	− .2	0
Perceived and evaluated health					
Perception of health (55)					
Excellent	5.3	.6	—[b]	− .4	0
Good	8.9	1.1	—	a	a
Fair	13.0	1.6	—	−1.4	a
Poor	14.5	1.4	—	− .8	0
Worry about health (73)					
A great deal	15.6	1.4	—	− .9	a
Some	9.2	1.4	—	.2	a
Hardly any	7.1	1.4	—	− .5	0
None	5.6	.5	—	− .4	a
Severity over all diagnoses (67)					
Elective only	7.9	1.5	—	−1.7	a
Elective and mandatory	7.3	.5	—	− .3	a
Mandatory only	13.4	1.3	—	− .6	0
Number of diagnoses (45)					
1	12.0	1.9	—	.1	a
2	8.3	.8	—	− .5	0
3	8.8	.9	—	−1.0	a
4 or more	11.5	1.3	—	− .9	a

Table 2.3. (Continued)

Characteristic	Hospital Days per Admission (33)				
	Social Survey Estimate	*Standard Error*	*Non-response Bias*	*Field Bias*	*Process-ing Bias*
Level of use					
Hospital days per admission (33)					
1 to 4	2.8	.1	—	a	0
5 to 10	6.9	.1	—	− .2	a
11 or more	30.4	3.0	—	− .8	a
Total	10.4	.7	.4	− .6	a

[a]Less than .05 days.

[b]Nonresponse biases were not calculated for perceived and evaluated health or level of use variables because necessary external data were not available.

Table 2.4. Differences between Groups by Unadjusted and Adjusted Social Survey Estimates for Hospital Admissions and Hospital Days

	Hospital Admissions per 100 Person-Years (30)			*Hospital Days per Admission (33)*		
Characteristic	*Unadjusted Difference*	*Adjusted Difference*	*Standard Error of the Difference*	*Unadjusted Difference*	*Adjusted Difference*	*Standard Error of the Difference*
Demographic						
Age of oldest family member (7)						
Less than 65 versus 65 or more	− 7.4**	− 6.2**	1.8	− 7.1**	− 7.2**	1.8
Family income (26)						
Nonpoor versus poor	− 2.4*	− 1.9*	1.3	− 2.1**	− 2.4**	1.4
Race (62)						
White versus nonwhite	3.2**	3.6**	1.1	− .1	− .4	1.1
Residence (64)						
SMSA other urban versus:						
Rural nonfarm	− .3	− .4	2.1	1.8	2.4	2.5
Rural farm	.2	.9	2.1	1.9	1.2	2.4
SMSA central city	2.2*	2.6*	2.0	1.6	1.2	2.3
Urban nonSMSA	− 1.6	− 1.2	2.1	4.6	4.3	2.4
Perceived and evaluated health[a]						
Perception of health (55)						
Excellent versus:						
Good	− 6.5**	− 6.4**	1.1	− 3.6**	− 4.0**	1.2
Fair	−15.7**	−14.2**	2.3	− 7.7**	− 6.7**	1.7
Poor	−39.9**	−36.7**	4.5	− 9.2**	− 8.8**	1.5
Worry about health (73)						
None versus:						
Hardly any	− 4.1**	− 3.7**	1.1	− 1.5*	− 1.4	1.5

Some	−17.0**	−16.4**	1.7	− 3.6*	− 4.2**	1.5
A great deal	−53.4**	−50.6**	4.1	−10.0**	− 9.5**	1.5
Severity over all diagnoses (67)						
Elective only versus:						
Elective and mandatory	−28.2**	−26.6**	2.1	.6	− .8	1.6
Mandatory only	−32.0**	−31.6**	2.1	− 5.5**	− 6.6**	2.0
Number of diagnoses (45)						
1 versus:						
2	− 5.9**	− 5.2**	1.8	3.7*	4.3**	2.1
3	−15.8**	−14.4**	2.8	3.2*	4.3**	2.1
4 or more	−39.7**	−38.8**	4.5	.5	1.5	2.3
Level of use[a]						
Hospital days per admission (33)						
1 to 4 versus:						
5 to 10	—	—	—	− 4.1**	− 3.9**	.1
11 or more	—	—	—	−27.6**	−26.8**	3.0

Note: * indicates that the difference shown is greater than one standard error; ** indicates that the difference shown is greater than two standard errors.

[a]Nonresponse biases were not calculated for these variables and are not a part of the adjusted figures.

Table 2.5. Comparison of Social Survey Estimates and Verification of Physician Visits for Persons Who Saw a Physician

	Percent of Population Seeing a Physician			
Type of Comparison	*Social Survey Equals*[a] *Verification*	*Social Survey Exceeds Verification*	*Verification Exceeds Social Survey*	*All Persons Seeing a Physician*
Difference in visits				
0	35.6%	—%	—%	35.6%
1	—	12.3	11.5	23.8
2 to 5	—	16.7	11.9	28.8
6 to 10	—	4.3	3.6	7.5
11 to 20	—	1.3	.8	2.2
21 or more	—	2.0	.6	2.5
All differences	35.6	36.6	27.8	100.0
Percent difference				
1.0% or less	35.6	—	—	35.6
1.1 to 5.0	—	.2	.1	.3
5.1 to 25.0	—	5.5	5.7	11.2
25.1 to 75.0	—	12.0	18.2	30.2
75.1 or more	—	17.5	2.0	19.5
Rejected in verification	—	1.5	—	1.5
Discovered in verification	—	—	1.8	1.8
All differences	35.6	36.6	27.8	100.0

Note: Persons for whom we received no response from a verification source for one or more of their visits were excluded.

[a]*Equals* means within one visit for the first category in this column and within one percentage point for the second.

Table 2.6. Error Components in Social Survey Estimates of Percent of Population Seeing a Physician

Characteristic	Percent of Population Seeing a Physician (55)				
	Social Survey Estimate	*Standard Error*	*Non-response Bias*	*Field Bias*	*Process-ing Bias*
Demographic					
Age of oldest family member (7)					
Less than 65 years	68%	1.1%	- .5%	-5.7%	—[a]
65 years or more	71	1.3	-1.3	-3.2	—
Family income (26)					
Nonpoor	72	1.0	-1.1	-5.3	—
Poor	59	1.5	- .9	-5.3	—
Race (62)					
White	70	1.1	-1.3	-5.0	—
Nonwhite	59	2.2	1.7	-8.8	—
Residence (64)					
Rural nonfarm	69	1.5	-1.6	-5.7	—
Rural farm	62	1.6	- .9	-6.9	—
SMSA central city	66	1.7	-1.3	-5.2	—
SMSA other urban	73	2.5	-1.2	-5.1	—
Urban nonSMSA	71	4.4	b	-4.5	—
Perceived Health					
Perception of health (55)					
Excellent	62	1.6	—[c]	-8.0	—
Good	68	1.4	—	-6.1	—
Fair	80	1.8	—	-6.0	—
Poor	90	1.4	—	-3.5	—
Worry about health (73)					
A great deal	92	1.6	—	-3.7	—
Some	87	1.0	—	-5.1	—
Hardly any	70	2.2	—	-7.8	—
None	57	1.4	—	-7.4	—
Total	67%	1.0%	- .8%	-5.3%	—

Note: The error components for the demographic characteristics and total are taken from Andersen (1975, p. 238).

[a]Fact of physician contact was not imputed.

[b]Less than .05 percent.

[c]Nonresponse biases were not calculated for perception of health and worry about health because necessary external data were not available.

Table 2.7. Error Components in Social Survey Estimates of Mean Number of Physician Visits for Persons Who Saw a Physician

	Visits for Persons Who Saw a Physician (59)				
Characteristic	*Social Survey Estimate*	*Standard Error*	*Non-response Bias*	*Field Bias*	*Process-ing Bias*
Demographic					
Age of oldest family member (7)					
Less than 65 years	5.4	.2	.1	− .7	a
65 years or more	7.8	.3	.1	−1.2	−.1
Family income (26)					
Nonpoor	5.5	.2	.1	− .7	a
Poor	6.5	.2	.1	−1.1	a
Race (62)					
White	5.7	.2	.1	− .7	.1
Nonwhite	6.0	.4	.1	−1.4	a
Residence (64)					
Rural nonfarm	5.4	.2	.1	− .5	.1
Rural farm	5.6	.6	.1	− .1	a
SMSA central city	6.0	.3	.1	−1.0	.1
SMSA other urban	5.5	.3	.1	− .6	a
Urban nonSMSA	6.1	.5	.1	−1.3	−.3
Perceived and evaluated health					
Perception of health (55)					
Excellent	3.6	.2	—[b]	.2	a
Good	5.3	.2	—	− .6	a
Fair	9.1	.5	—	−2.4	a
Poor	14.2	1.1	—	−4.3	−.7
Worry about health (73)					
A great deal	11.7	.7	—	−3.1	−.1
Some	7.4	.3	—	−1.4	.1
Hardly any	4.9	.2	—	− .4	a
None	3.6	.2	—	− .1	.1
Severity over all diagnoses (67)					
Elective only	3.2	.2	—	− .8	a
Elective and mandatory	8.3	.2	—	−1.1	−.1
Mandatory only	7.2	.3	—	−1.4	a
Number of diagnoses (45)					
1	3.8	.2	—	−1.0	a
2	5.6	.3	—	− .6	−.1
3	8.0	.4	—	−1.5	.2
4 or more	11.6	.3	—	−1.5	−.1

Table 2.7. (Continued)

Characteristic	Visits for Persons Who Saw a Physician (59)				
	Social Survey Estimate	*Standard Error*	*Non-response Bias*	*Field Bias*	*Process-ing Bias*
Level of use					
Physician visits per person (59)					
1	1.0	0	—	– .6	.1
2 to 4	2.7	a	—	– .5	a
5 or more	12.1	.3	—	–3.6	–.1
Total	5.7	.2	a	– .8	a

Note: Only persons who saw a physician are included in this table. Andersen (1975) shows error components for physician visits per capita.

[a]Less than .05 visits.

[b]Nonresponse biases were not calculated for perceived and evaluated health or level of use variables because necessary external data were not available.

Table 2.8. Differences between Groups by Unadjusted and Adjusted Social Survey Estimates for Physician Contact and Physician Visits

Characteristic	*Percent of Population Seeing a Physician (55)*			*Visits for Persons Who Saw a Physician (59)*		
	Unadjusted Difference	*Adjusted Difference*	*Standard Error of the Difference*	*Unadjusted Difference*	*Adjusted Difference*	*Standard Error of the Difference*
Demographic						
Age of oldest family member (7)						
Less than 65 versus 65 or more	– 3.0*	– 4.7**	1.7	– 2.4**	–1.9**	.4
Family income (26)						
Nonpoor versus poor	13.0**	12.8**	1.8	– 1.0**	– .6**	.3
Race (62)						
White versus nonwhite	11.0**	11.8**	2.4	– .3	.2	.4
Residence (64)						
SMSA other urban versus:						
Rural nonfarm	4.0*	5.0*	2.9	.1	– .1	.4
Rural farm	11.0**	12.5**	3.0	– .1	– .7	.7
SMSA central city	7.0**	7.1**	3.0	– .5*	– .3	.4
Urban nonSMSA	2.0	1.8	5.1	– .6*	.2	.6
Perceived and evaluated health[a]						
Perception of health (55)						
Excellent versus:						
Good	– 6.2**	– 8.1**	2.1	– 1.7**	– .9**	.3
Fair	–17.7**	–19.7**	2.4	– 5.5**	–3.0**	.5
Poor	–28.0**	–32.5**	2.1	–10.6**	–5.6**	1.1
Worry about health (73)						
None versus:						
Hardly any	–13.1**	–12.7**	2.6	– 1.3**	–1.1**	.3
Some	–30.0**	–32.3**	1.7	– 3.8**	–2.6**	.4
A great deal	–35.0**	–38.7**	2.1	– 8.1**	–5.4**	.7

Severity over all diagnoses (67)[b]						
Elective only versus:						
Elective and mandatory	—	—	—	− 5.1**	−4.7**	.3
Mandatory only	—	—	—	− 4.0**	−3.5**	.4
Number of diagnoses (45)[b]						
1 versus:						
2	—	—	—	− 1.8**	−2.1**	.4
3	—	—	—	− 4.2**	−3.9**	.4
4 or more	—	—	—	− 7.8**	−7.2**	.4
Level of use[a]						
Physician visits per person (59)[b]						
1 versus:						
2 to 4	—	—	—	− 1.7**	−1.7**	c
5 or more	—	—	—	−11.1**	−7.9**	.3

Note: * indicates that the difference shown is greater than one standard error; ** indicates that the difference shown is greater than two standard errors.

[a] Nonresponse biases were not calculated for these variables and are not a part of the adjusted figures.

[b] These variables are not applicable to physician contact.

[c] Less than .05 visits.

Table 3.1. Comparison of Social Survey Estimates and Verification of Expenditures per Hospital Admission

Type of Comparison	Percent of All Admissions			
	Social Survey Equals[a] *Verification*	*Social Survey Exceeds Verification*	*Verification Exceeds Social Survey*	*All Admissions*
Dollar difference				
$1 or less	21.8%	—%	—%	21.8%
2 to 25	—	9.2	7.4	16.6
26 to 100	—	10.5	11.4	21.9
101 to 200	—	7.2	8.0	15.2
201 to 400	—	7.0	5.6	12.6
401 or more	—	6.1	6.0	12.1
All differences	21.8	40.0	38.4	100.0
Percent difference				
1.0% or less	23.3	—	—	23.3
1.1 to 5.0	—	6.3	6.9	13.2
5.1 to 25.0	—	10.6	12.7	23.3
25.1 to 75.0	—	9.3	12.9	22.2
75.1 or more	—	5.5	.6	6.1
Rejected in verification	—	6.9	—	6.9
Discovered in verification	—	—	4.8	4.8
All differences	23.3	38.6	37.9	100.0

Note: Admissions for which we received no response from any verification source were excluded.

[a] *Equals* means within one dollar for the first category in this column and within one percentage point for the second.

Table 3.2. Error Components in Social Survey Estimates of Hospital Expenditures per Admission

	Expenditures per Admission (23)				
Characteristic	*Social Survey Estimate*	*Standard Error*	*Non-response Bias*[a]	*Field Bias*[a]	*Process-ing Bias*[a]
Demographic					
Age of oldest family member (7)[b]					
Less than 65 years	$ 640	$ 47	$22	$- 4	$-11
65 years or more	863	58	25	21	9
Family income (26)					
Nonpoor	674	44	23	4	- 7
Poor	715	55	22	- 14	- 7
Race (62)					
White	684	40	23	- 4	- 2
Nonwhite	688	82	23	33	-62
Residence (64)					
Rural nonfarm	557	66	16	- 21	- 6
Rural farm	575	63	11	- 6	- 8
SMSA central city	727	41	29	- 6	17
SMSA other urban	910	150	31	23	-31
Urban nonSMSA	444	33	15	- 7	6
Perceived and evaluated health					
Perception of health (55)					
Excellent	488	84	—[c]	- 29	18
Good	609	68	—	25	6
Fair	698	59	—	- 6	-22
Poor	867	64	—	- 21	-47
Worry about health (73)					
A great deal	1,036	82	—	3	-47
Some	560	67	—	9	20
Hardly any	442	65	—	26	9
None	441	35	—	- 15	10
Severity over all diagnoses (67)					
Elective only	498	73	—	- 72	2
Elective and mandatory	550	46	—	- 12	19
Mandatory only	830	61	—	12	-35
Number of diagnoses (45)					
1	626	84	—	16	7
2	573	51	—	1	-12
3	563	46	—	- 18	-30
4 or more	883	82	—	155	d

Table 3.2. (Continued)

Characteristic	Expenditures per Admission (23)				
	Social Survey Estimate	Standard Error	Non-response Bias[a]	Field Bias[a]	Processing Bias[a]
Level of use					
Expenditures per admission (23)					
$0 to 300	184	5	—	19	21
301 to 600	426	5	—	13	13
601 or more	1,556	112	—	− 30	−66
Total	$ 685	$ 39	$23	$− d	$− 7

Note: An earlier version of the above data appeared in Andersen, Kasper, and Frankel (1976). Slight differences in the numbers presented here are due to refinements in our methods of calculating bias and adjustments to three high-weight cases in our data set.

[a]In this and all subsequent tables in the chapter, a positive bias figure indicates that the amount should be added to the social survey estimate to reach the adjusted estimate and a negative figure indicates that the amount should be subtracted.

[b]In this and all subsequent tables in the chapter, the numbers in parentheses refer to the variable definition list found in Appendix C.

[c]Nonresponse biases were not calculated for perceived and evaluated health or level of use variables because necessary external data were not available.

[d]Less than fifty cents.

Table 3.3. Comparison of Social Survey Estimates and Verification of Expenditures for Persons with Physician Office Visits

Type of Comparison	*Percent of Persons with Visits*			
	Social Survey Equals[a] Verification	*Social Survey Exceeds Verification*	*Verification Exceeds Social Survey*	*All Persons with Physician Office Visits*
Dollar difference				
$1 or less	18.9%	—%	—%	18.9%
2 to 25	—	27.2	25.6	52.8
26 to 100	—	11.0	11.2	22.2
101 to 200	—	2.2	2.2	4.4
201 to 400	—	.6	.6	1.2
401 or more	—	.3	.2	.5
All differences	18.9	41.3	39.8	100.0
Percent difference				
1.0% or less	15.5	—	—	15.5
1.1 to 5.0	—	1.0	1.4	2.4
5.1 to 25.0	—	7.9	8.3	16.2
25.1 to 75.0	—	12.2	21.3	33.5
75.1 or more	—	18.2	4.6	22.8
Rejected in verification	—	3.3	—	3.3
Discovered in verification	—	—	6.1	6.1
All differences	15.5	42.6	41.7	100.0

Note: Persons for whom we received no response from a verification source for one or more of their visits were excluded.

[a]*Equals* means within one dollar for the first category in this column and within one percentage point for the second.

Table 3.4. Error Components in Social Survey Estimates of Expenditures for Persons with Physician Office Visits

	Office Visit Expenditures per Person (24)				
Characteristic	*Social Survey Estimate*	*Standard Error*	*Non-response Bias*	*Field Bias*	*Process-ing Bias*
Demographic					
Age of oldest family member (7)					
Less than 65 years	$ 50	$ 2	$a	$- 3	$- 2
65 years or more	67	3	a	1	- 4
Family income (26)					
Nonpoor	52	2	a	- 3	- 2
Poor	55	5	a	1	- 4
Race (62)					
White	53	2	a	- 2	- 2
Nonwhite	54	6	3	- 5	- 4
Residence (64)					
Rural nonfarm	44	3	a	- 1	a
Rural farm	41	4	a	2	- 1
SMSA central city	64	5	a	- 1	- 8
SMSA other urban	56	4	a	- 2	a
Urban nonSMSA	46	5	1	- 6	- 2
Perceived and evaluated health					
Perception of health (55)					
Excellent	36	3	—[b]	2	a
Good	47	2	—	a	- 1
Fair	82	8	—	-11	1
Poor	136	15	—	-20	-25
Worry about health (73)					
A great deal	110	9	—	-15	-12
Some	71	5	—	- 3	- 1
Hardly any	43	2	—	5	- 2
None	33	2	—	- 2	a
Severity over all diagnoses (67)					
Elective only	32	2	—	- 7	a
Elective and mandatory	72	3	—	- 2	- 2
Mandatory only	65	5	—	- 5	- 5
Number of diagnoses (45)					
1	33	2	—	- 6	- 1
2	52	3	—	- 7	a
3	67	4	—	- 1	- 1
4 or more	112	6	—	a	-12

Table 3.4. (Continued)

Characteristic	Office Visit Expenditures per Person (24)				
	Social Survey Estimate	*Standard Error*	*Non-response Bias*	*Field Bias*	*Process-ing Bias*
Level of use					
Expenditures for persons with physician office visits (24)					
$1 to 15	10	a	—	10	1
16 to 45	28	a	—	7	1
46 or more	128	4	—	−30	− 9
Total	$ 53	$ 2	$a	$− 2	$− 2

[a]Less than fifty cents.

[b]Nonresponse biases were not calculated for perceived and evaluated health or level of use variables because necessary external data were not available.

Table 3.5. Comparison of Social Survey Estimates and Verification of Expenditures for Persons with Emergency Room, Outpatient, or Clinic Visits

Type of Comparison	*Percent of Persons with Visits*			
	Social Survey Equals[a] *Verification*	*Social Survey Exceeds Verification*	*Verification Exceeds Social Survey*	*All Persons with Emergency Room, Outpatient, or Clinic Visits*
Dollar difference				
$1 or less	23.1%	—%	—%	23.1%
2 to 25	—	20.0	19.7	39.7
26 to 100	—	13.4	8.3	21.7
101 to 200	—	5.5	2.0	7.5
201 to 400	—	7.1	.6	7.7
401 or more	—	.3	b	.3
All differences	23.1	46.3	30.6	100.0
Percent difference				
1.0% or less	22.0	—	—	22.0
1.1 to 5.0	—	.2	b	.2
5.1 to 25.0	—	1.8	2.9	4.7
25.1 to 75.0	—	2.9	8.9	11.8
75.1 or more	—	8.5	2.5	11.0
Rejected in verification	—	33.0	—	33.0
Discovered in verification	—	—	17.2	17.2
All differences	22.0	46.4	31.5	100.0

Note: Persons for whom we received no response from a verification source for one or more of their visits were excluded.

[a]*Equals* means within one dollar for the first category in this column and within one percentage point for the second.

[b]Less than .05 percent.

Table 3.6. Error Components in Social Survey Estimates of Expenditures for Persons with Emergency Room, Outpatient, or Clinic Visits

Characteristic	*Emergency Room, Outpatient, or Clinic Visit Expenditures per Person (14)*				
	Social Survey Estimate	*Standard Error*	*Non-response Bias*	*Field Bias*	*Processing Bias*
Demographic					
Age of oldest family member (7)					
Less than 65 years	$ 43	$ 4	$ 1	$- 3	$24
65 years or more	50	7	-1	- 1	a
Family income (26)					
Nonpoor	39	4	b	- 4	13
Poor	57	9	b	6	28
Race (62)					
White	42	4	-1	- 4	16
Nonwhite	52	10	4	5	18
Residence (64)					
Rural nonfarm	36	4	-1	-10	a
Rural farm	28	4	-1	16	a
SMSA central city	61	9	b	4	29
SMSA other urban	36	4	b	- 4	a
Urban nonSMSA	27	5	3	9	a
Perceived and evaluated health					
Perception of health (55)					
Excellent	30	3	—[c]	- 9	10
Good	42	6	—	- 4	13
Fair	66	15	—	9	a
Poor	75	9	—	35	a
Worry about health (73)					
A great deal	74	14	—	4	42
Some	62	11	—	-18	24
Hardly any	36	3	—	- 5	4
None	25	2	—	b	13
Severity over all diagnoses (67)					
Elective only	24	3	—	- 4	a
Elective and mandatory	56	8	—	-14	16
Mandatory only	51	5	—	4	36
Number of diagnoses (45)					
1	34	4	—	3	5
2	46	8	—	-11	16
3	55	7	—	-11	16
4 or more	59	11	—	- 6	29

Table 3.6. (Continued)

Characteristic	Emergency Room, Outpatient, or Clinic Visit Expenditures per Person (14)				
	Social Survey Estimate	*Standard Error*	*Non-response Bias*	*Field Bias*	*Process-ing Bias*
Level of use					
Expenditures for persons with emergency room, outpatient, or clinic visits (14)					
$1 to 10	7	b	—	8	a
11 to 30	18	b	—	6	20
31 or more	105	9	—	−19	16
Total	$ 44	$ 4	$ b	$− 3	$17

[a]The unweighted N's were too small (fewer than 25 persons) in e_s or e_v (Section IV in Appendix A) to calculate a bias.

[b]Less than fifty cents.

[c]Nonresponse biases were not calculated for perceived and evaluated health or level of use variables because necessary external data were not available.

Table 3.7. Differences between Groups by Unadjusted and Adjusted Social Survey Estimates of Expenditures for Hospital Admissions

Characteristic	Expenditures per Admission (23)		
	Unadjusted Difference	Adjusted Difference	Standard Error of the Difference
Demographic			
Age of oldest family member (7)			
Less than 65 versus 65 or more	$- 222**	$- 271**	$ 74.6
Family income (26)			
Nonpoor versus poor	- 41	- 22	70.4
Race (62)			
White versus nonwhite	- 4	19	91.2
Residence (64)			
SMSA other urban versus:			
Rural nonfarm	353**	387**	163.9
Rural farm	335**	361**	162.7
SMSA central city	183*	166*	155.5
Urban nonSMSA	467**	475**	159.1
Perceived and evaluated health[a]			
Perception of health (55)			
Excellent versus:			
Good	- 121*	- 163*	108.1
Fair	- 210**	- 193*	102.6
Poor	- 379**	- 322**	105.6
Worry about health (73)			
None versus:			
Hardly any	- 1	- 41	73.8
Some	- 119*	- 153**	75.6
A great deal	- 595**	- 556**	89.2
Severity over all diagnoses (67)			
Elective only versus:			
Elective and mandatory	- 52	- 129*	86.3
Mandatory only	- 332**	- 379**	95.1
Number of diagnoses (45)			
1 versus:			
2	53	87	98.3
3	63	134*	95.8
4 or more	- 257**	- 389**	117.4

Table 3.7. (Continued)

Characteristic	Expenditures per Admission (23)		
	Unadjusted Difference	Adjusted Difference	Standard Error of the Difference
Level of use[a]			
Expenditures per admission (23)			
$0 to 300 versus:			
301 to 600	− 242**	− 229**	7.1
601 or more	−1,371**	−1,236**	112.1

Note: * indicates that the difference shown is greater than one standard error; ** indicates that the difference shown is greater than two standard errors.

[a]Nonresponse biases were not calculated for these variables and are not a part of the adjusted figures.

Table 3.8. Differences between Groups by Unadjusted and Adjusted Social Survey Estimates of Expenditures for Physician Office Visits and for Emergency Room, Outpatient, and Clinic Visits

	Office Visit Expenditures per Person (24)			*Emergency Room, Outpatient, or Clinic Visit Expenditures per Person (14)*		
Characteristic	*Unadjusted Difference*	*Adjusted Difference*	*Standard Error of the Difference*	*Unadjusted Difference*	*Adjusted Difference*	*Standard Error of the Difference*
Demographic						
Age of oldest family member (7)						
Less than 65 versus 65 or more	$– 17**	$–19**	$ 3.6	$– 7	$ 17**	$ 8.1
Family income (26)						
Nonpoor versus poor	– 3	– 5	5.4	–18*	–43**	9.8
Race (62)						
White versus nonwhite	– 1	1	6.3	–10	–26**	10.8
Residence (64)						
SMSA other urban versus:						
Rural nonfarm	12**	11**	5.0	0	7*	5.7
Rural farm	15**	12**	5.7	33**	–11*	5.7
SMSA central city	– 8*	– 1	6.4	–25**	–62**	9.8
Urban nonSMSA	10*	15**	6.4	9*	– 7**	6.4
Perceived and evaluated health[a]						
Perception of health (55)						
Excellent versus:						
Good	– 11**	– 8**	3.6	–12*	–20**	6.7
Fair	– 46**	–35**	8.5	–36**	–39**	15.3
Poor	–100**	–53**	15.3	–45**	–74**	9.5
Worry about health (73)						
None versus:						
Hardly any	– 10**	–15**	2.8	–11**	3	3.6
Some	– 38**	–36**	5.4	–37**	–30**	11.2
A great deal	– 77**	–52**	9.2	–49**	–82**	14.1

Table 3.8. (Continued)

Characteristic	Office Visit Expenditures per Person (24)			Emergency Room, Outpatient, or Clinic Visit Expenditures per Person (14)		
	Unadjusted Difference	Adjusted Difference	Standard Error of the Difference	Unadjusted Difference	Adjusted Difference	Standard Error of the Difference
Severity over all diagnoses (67)						
Elective only versus:						
Elective and mandatory	− 40**	−43**	3.6	−32**	−38**	8.5
Mandatory only	− 33**	−30**	5.4	−27**	−71**	9.4
Number of diagnoses (45)						
1 versus:						
2	− 19**	−19**	3.6	−12*	− 9*	8.9
3	− 34**	−39**	4.5	−21**	−18**	8.1
4 or more	− 79**	−74**	7.3	−25**	−40**	11.7
Level of use[a]						
Expenditures for persons with physician visits (24)						
$1 to 15 versus:						
16 to 45	− 18**	−15**	b	—	—	—
46 or more	−118**	−68**	.4	—	—	—
Expenditures for persons with emergency room, outpatient, or clinic visits (14)						
$1 to 10 versus:						
11 to 30	—	—	—	−11**	−29**	b
31 or more	—	—	—	−98**	−87**	.9

Note: * indicates that the difference shown is greater than one standard error; ** indicates that the difference shown is greater than two standard errors.

[a]Nonresponse biases were not calculated for these variables and are not a part of the adjusted figures.

[b]Less than fifty cents.

Table 4.1. Comparison of Social Survey Estimates and Verification of Expenditures for Premiums per Policy

Type of Comparison	*Percent of All Policies*			
	Social Survey Equals[a] Verification	*Social Survey Exceeds Verification*	*Verification Exceeds Social Survey*	*All Policies*
Dollar difference				
$1 or less	16.4%	—%	—%	16.4%
2 to 25	—	15.6	10.9	26.5
26 to 100	—	19.3	11.6	30.9
101 to 200	—	9.3	7.4	16.7
201 to 400	—	4.5	2.7	7.2
401 or more	—	1.0	b	1.3
All differences	16.4	49.7	32.9	100.0
Percent difference				
1.0% or less	14.9	—	—	14.9
1.1 to 5.0	—	5.2	3.0	8.2
5.1 to 25.0	—	11.3	8.5	19.8
25.1 to 75.0	—	9.3	10.8	21.1
75.1 or more	—	10.4	3.0	13.4
Policy rejected	—	5.6	—	5.6
Policy discovered	—	—	1.3	1.3
Premium rejected	—	9.1	—	9.1
Premium discovered	—	—	7.5	7.5
All differences	14.9	50.9	34.1	100.0

Note: Policies for which we received no response from any verification source were excluded, and those with "no answer" or "don't know" in social survey or verification were excluded, as were 332 (unweighted) cases where both verification and social survey reported "$0" for family-paid premiums. Also excluded were 20 policies rejected and 9 discovered in verification where the reporting source claimed no family expenditure.

[a] *Equals* means within one dollar for the first category in this column and within one percentage point for the second.

[b] Less than .05 percent.

Table 4.2. Error Components in Social Survey Estimates of Premiums per Health Insurance Policy

Characteristic	*Premium per Policy*[a] *(61)*		
	Social Survey Estimate	*Standard Error*	*Field Bias*[b]
Demographic			
Age of oldest family member (7)[c]			
Less than 65 years	$106	$ 5	$- 18
65 years or more	96	3	- 6
Family income (26)			
Nonpoor	103	5	- 16
Poor	108	5	- 12
Race (62)			
White	104	4	- 16
Nonwhite	94	9	- 7
Residence (64)			
Rural nonfarm	106	7	- 19
Rural farm	144	13	- 8
SMSA central city	87	5	- 12
SMSA other urban	110	11	- 21
Urban nonSMSA	109	9	- 11
Other			
Premium for health insurance policy (61)[d]			
Low ($1 to 76)	46	1	8
Medium ($77 to 169)	120	1	- 18
High ($170 or more)	304	7	-120
Total family expenditures for health services (17)			
Low ($0 to 202)	93	6	- 17
Medium ($203 to 656)	100	6	- 13
High ($657 or more)	116	8	- 18
Enrollment status of policy (13)			
Work group			
Employer pays none	220	16	-107
Employer pays some	142	6	- 40
Employer pays all	0	0	- 18
Not work group	156	7	- 16
Total	$104	$ 4	$- 16

[a]Responses of "don't know" and "no answer" are omitted.

[b]In this and all subsequent tables in the chapter, a positive bias figure indicates that the amount should be added to the social survey estimate to reach the adjusted estimate, and a negative figure indicates the amount should be subtacted.

[c]In this and all subsequent tables in the chapter, the numbers in parentheses refer to the variable definition list found in Appendix C.

[d]Policies discovered in verification are omitted.

Table 4.3. Comparison of Social Survey and Verification Responses for Health Insurance Coverage per Policy by Types of Coverage

Type of Coverage	*Percent Coverage Claimed*				
	Social Survey: Yes Verification: Yes	*Social Survey: Yes Verification: No*	*Social Survey: No Verification: Yes*	*Social Survey: No Verification: No*	*Total Policies*
Hospital (40)	83%	9%	4%	4%	100%
Surgical-medical (43)	76	14	4	6	100
Physician visit (42)	3	18	9	70	100
Dental (38)	4	5	7	84	100
Drug (39)	1	7	3	90	100
Major medical (41)	42	26	7	26	100

Note: Only policies for which verification and social survey information was available are included; unverified, discovered, or rejected policies are excluded. Further, policies for which the social survey or verification response was "don't know" or "no answer" are excluded.

Table 4.4. Error Components in Social Survey Estimates of Percent of Policies with Hospital Coverage

	Percent of Policies with with Hospital Coverage (40)		
Characteristic	*Social Survey Estimate*	*Standard Error*	*Field Bias*
Demographic			
Age of oldest family member (7)			
Less than 65 years	90%	1.0%	−5.1%
65 years or more	93	1.1	− .9
Family income (26)			
Nonpoor	90	1.0	−4.6
Poor	93	1.1	−1.4
Race (62)			
White	90	.9	−4.3
Nonwhite	94	2.2	−3.1
Residence (64)			
Rural nonfarm	90	1.6	−6.9
Rural farm	89	2.1	−4.1
SMSA central city	91	1.4	.7
SMSA other urban	92	1.4	−4.1
Urban nonSMSA	90	3.4	−2.4
Other			
Family hospital admissions (28)			
Low (0 admissions)	90	1.0	−5.7
Medium (1 admission)	90	1.7	−1.6
High (2 or more)	92	2.5	.1
Enrollment status of policy (13)[a]			
Work group			
Employer pays none	92	2.3	−2.4
Employer pays some	93	1.5	−6.8
Employer pays all	91	1.3	−6.9
Not work group	89	1.2	.9
Total	90%	.8%	−4.2%

[a]Policies discovered in verification are omitted.

Table 4.5. Error Components in Social Survey Estimates of Percent of Policies with Surgical-Medical Coverage

Characteristic	Percent of Policies with Surgical-Medical Coverage (43)		
	Social Survey Estimate	*Standard Error*	*Field Bias*
Demographic			
Age of oldest family member (7)			
Less than 65 years	88%	1.1%	− 7.9%
65 years or more	89	1.5	−16.7
Family income (26)			
Nonpoor	88	1.0	− 9.2
Poor	87	1.5	−12.9
Race (62)			
White	88	1.0	− 9.7
Nonwhite	90	2.7	− 9.0
Residence (64)			
Rural nonfarm	88	1.5	−11.8
Rural farm	88	1.8	−10.2
SMSA central city	87	3.0	− 8.1
SMSA other urban	89	1.7	− 7.0
Urban nonSMSA	85	3.6	−10.2
Other			
Family expenditures for hospital services (18)			
None ($0)	88	1.2	−11.2
Low ($1 to 386)	89	2.1	− 8.5
Middle ($387 to 806)	87	3.0	− 5.2
High ($807 or more)	88	4.1	− 5.5
Enrollment status of policy (13)[a]			
Work group			
Employer pays none	91	2.4	− 4.2
Employer pays some	92	1.4	− 8.5
Employer pays all	91	1.5	− 9.1
Not work group	82	1.6	−11.4
Total	88%	1.0%	− 9.6%

[a]Policies discovered in verification are omitted.

Table 4.6. Error Components in Social Survey Estimates of Percent of Policies with Major Medical Coverage

	Percent of Policies with Major Medical Coverage (41)		
Characteristic	*Social Survey Estimate*	*Standard Error*	*Field Bias*
Demographic			
Age of oldest family member (7)			
Less than 65 years	67%	1.7%	−13.6%
65 years or more	57	3.0	−41.3
Family income (26)			
Nonpoor	66	1.8	−17.2
Poor	58	3.3	−30.4
Race (62)			
White	65	1.7	−18.4
Nonwhite	68	5.8	−21.0
Residence (64)			
Rural nonfarm	62	4.7	−14.2
Rural farm	70	2.6	−24.2
SMSA central city	65	3.2	−17.3
SMSA other urban	66	3.6	−18.5
Urban nonSMSA	56	4.9	−13.7
Other			
Total family expenditures for health services (17)			
Low ($0 to 202)	66	2.9	−17.5
Medium ($203 to 656)	63	2.6	−19.6
High ($657 or more)		3.0	−18.5
Enrollment status of policy (13)[a]			
Work group			
Employer pays none	72	4.9	−18.8
Employer pays some	80	2.4	− 6.2
Employer pays all	67	2.0	−14.0
Not work group	50	2.4	−35.5
Total	65%	1.7%	−18.5%

[a]Policies discovered in verification are omitted.

Table 4.7. Error Components in Social Survey Estimates of Percent of Policies with Physician Visit Coverage

Characteristic	*Percent of Policies with Physician Visit Coverage (42)*		
	Social Survey Estimate	*Standard Error*	*Field Bias*
Demographic			
Age of oldest family member (7)			
Less than 65 years	20%	1.8%	−13.1%
65 years or more	19	2.0	7.0
Family income (26)			
Nonpoor	20	1.9	−10.2
Poor	20	2.6	1.2
Race (62)			
White	20	1.8	− 8.3
Nonwhite	22	2.8	−12.0
Residence (64)			
Rural nonfarm	26	3.5	−18.6
Rural farm	15	1.9	1.3
SMSA central city	11	3.0	2.9
SMSA other urban	23	2.5	−11.6
Urban nonSMSA	23	4.2	−11.7
Other			
Total family physician visits (60)			
None (0)	14	3.5	− 6.8
Low (1 to 5)	20	2.2	− 9.2
Medium (6 to 14)	21	2.7	− 5.5
High (15 or more)	21	3.0	−11.8
Enrollment status of policy (13)[a]			
Work group			
Employer pays none	12	3.6	− 2.4
Employer pays some	22	3.2	−16.9
Employer pays all	26	3.1	−13.5
Not work group	15	1.6	.5
Total	20%	1.7%	− 8.5%

[a]Policies discovered in verification are omitted.

Table 4.8. Error Components of Social Survey Estimates of Percent of Policies with Drug Coverage

Characteristic	*Percent of Policies with Drug Coverage (39)*		
	Social Survey Estimate	*Standard Error*	*Field Bias*
Demographic			
Age of oldest family member (7)			
Less than 65 years	9%	1.2%	−6.5%
65 years or more	5	1.2	4.5
Family income (26)			
Nonpoor	8	1.2	−4.7
Poor	7	1.4	1.0
Race (62)			
White	8	1.2	−3.8
Nonwhite	7	1.6	−5.3
Residence (64)			
Rural nonfarm	10	1.8	−6.0
Rural farm	6	1.3	− .4
SMSA central city	5	1.6	−2.2
SMSA other urban	10	2.6	−6.2
Urban nonSMSA	4	1.9	−1.3
Other			
Total family expenditures for drugs (16)			
Low ($0 to 26)	7	1.3	−3.3
Medium ($27 to 92)	7	1.4	−4.4
High ($93 or more)	9	2.5	−3.8
Enrollment status of policy (13)[a]			
Work group			
Employer pays none	4	1.6	−3.0
Employer pays some	11	2.3	−9.5
Employer pays all	12	2.4	−6.6
Not work group	3	.6	2.6
Total	8%	1.1%	−3.9%

[a]Policies discovered in verification are omitted.

Table 4.9. Error Components in Social Survey Estimates of Percent of Policies with Dental Coverage

	Percent of Policies with Dental Coverage (38)		
Characteristic	*Social Survey Estimate*	*Standard Error*	*Field Bias*
Demographic			
Age of oldest family member (7)			
Less than 65 years	11%	1.2%	1.5%
65 years or more	3	.8	1.9
Family income (26)			
Nonpoor	10	1.1	1.5
Poor	5	1.2	3.5
Race (62)			
White	9	1.0	2.1
Nonwhite	11	2.3	-5.7
Residence (64)			
Rural nonfarm	10	1.4	- .5
Rural farm	11	2.0	1.7
SMSA central city	6	2.0	3.2
SMSA other urban	9	1.3	2.0
Urban nonSMSA	7	3.2	5.5
Other			
Family expenditures for dental services (15)			
None ($0)	7	1.1	2.4
Low ($1 to 37)	8	1.6	2.3
Medium ($38 to 119)	12	2.3	-1.5
High ($120 or more)	12	1.9	4.7
Enrollment status of policy (13)[a]			
Work group			
Employer pays none	9	3.0	1.5
Employer pays some	10	1.5	6.4
Employer pays all	17	2.3	- .7
Not work group	3	.6	.1
Total	10%	1.0%	1.6%

[a]Policies discovered in verification are omitted.

Table 4.10. Mean Payment for Hospital Admissions by Type of Payer and Source of Information

Payer	Source of Information				
	Discovered in Verification	Rejected in Verification	Reported in Social Survey and Verified[c]		Reported in Social Survey, No Verification[a]
			Social Survey Mean	Verification Mean	
Out-of-Pocket	$ 12	$ 35	$112	$ 74	$146
Insurance	247	150	314	360	349
Medicare	93	138	107	114	159
Welfare	66	143	73	76	45
Total[b]	$453 (4%)[c]	$490 (6%)[c]	$708	$703	$592 (7%)[c]
			(82%)[c]		

[a]For out-of-pocket and insurance payers, cases suspected of having received indemnity payments were shifted from "Reported in Social Survey and Verified" to "Reported in Social Survey, No Verification" because of incomplete reporting of such payments in verification.

[b]Includes other payers not noted above, for example, the Veterans Administration.

[c]Weighted percent of all admissions reported in social survey and discovered in verification.

Table 4.11. Error Components in Social Survey Estimates of Out-of-Pocket Payments per Hospital Admission

Characteristic	Out-of-Pocket Payments per Admission (53)			
	Social Survey Estimate	*Standard Error*	*Field Bias*	*Processing Bias*
Demographic				
Age of oldest family member (7)				
Less than 65 years	$126	$28	$-41	$-3
65 years or more	58	9	- 7	1
Family income (26)				
Nonpoor	120	31	-34	-2
Poor	90	16	-33	-7
Race (62)				
White	120	26	-36	-3
Nonwhite	42	9	-21	-2
Residence (64)				
Rural nonfarm	83	15	-28	-9
Rural farm	99	30	-21	-2
SMSA central city	72	16	-32	-2
SMSA other urban	197	74	-42	-1
Urban nonSMSA	78	13	-35	1
Level of use				
Expenditures per admission (23)[a]				
Low ($0 to 300)	29	3	- 8	b
Medium ($301 to 600)	86	8	-23	-1
High ($601 or more)	238	77	-70	-8
Total	$112	$23	$-34	$-3

Note: An earlier version of the above data appeared in Andersen, Kasper, and Frankel (1976). Slight differences in the numbers presented here (and in Tables 4.12 to 4.16) are due to refinements in our methods of calculating bias and adjustments to three high-weight cases in our data.

[a]Admissions discovered in verification are omitted.

[b]Less than fifty cents.

Table 4.12. Error Components in Social Survey Estimates of Insurance Payments per Hospital Admission

Characteristic	*Insurance Payments per Admission (51)*			
	Social Survey Estimate	*Standard Error*	*Field Bias*	*Process-ing Bias*
Demographic				
Age of oldest family member (7)				
Less than 65 years	$389	$ 39	$ 5	$ 11
65 years or more	127	19	2	4
Family income (26)				
Nonpoor	397	32	5	15
Poor	165	42	4	− 4
Race (62)				
White	346	29	2	13
Nonwhite	257	88	38	−17
Residence (64)				
Rural nonfarm	297	61	− 1	7
Rural farm	210	27	−19	3
SMSA central city	333	47	5	28
SMSA other urban	478	109	12	6
Urban nonSMSA	201	32	12	− 2
Level of use				
Expenditures per admission (23)[a]				
Low ($0 to 300)	114	6	12	6
Medium ($301 to 600)	229	13	16	10
High ($601 or more)	717	86	− 9	12
Total	$337	$ 30	$ 6	$ 10

[a]Admissions discovered in verification are omitted.

Table 4.13. Error Components in Social Survey Estimates of Medicare Payments per Hospital Admission

	Medicare Payments per Admission (52)			
Characteristic	*Social Survey Estimate*	*Standard Error*	*Field Bias*	*Processing Bias*
Demographic				
Age of oldest family member (7)				
Less than 65 years	$ —[a]	$—	$ —	$ —
65 years or more	567	54	31	12
Family income (26)				
Nonpoor	82	18	− 1	1
Poor	202	33	16	5
Race (62)				
White	115	15	b	3
Nonwhite	91	20	21	− 1
Residence (64)				
Rural nonfarm	104	22	−10	3
Rural farm	186	40	− 1	− 6
SMSA central city	115	22	15	11
SMSA other urban	122	39	− 2	− 1
Urban nonSMSA	74	22	15	b
Level of use				
Expenditures per admission (23)[c]				
Low ($0 to 300)	13	2	2	5
Medium ($301 to 600)	50	7	8	− 1
High ($601 or more)	300	42	− 6	3
Total	$113	$14	$ 3	$ 2

[a]Not applicable.

[b]Less than fifty cents.

[c]Admissions discovered in verification are omitted.

Table 4.14. Error Components in Social Survey Estimates of Welfare Payments per Admission

Characteristic	Welfare Payments per Admission (54)			
	Social Survey Estimate	*Standard Error*	*Field Bias*	*Processing Bias*
Demographic				
Age of oldest family member (7)				
Less than 65 years	$ 80	$15	$ 5	$ − 3
65 years or more	57	18	4	−27
Family income (26)				
Nonpoor	27	7	12	− 4
Poor	216	40	−18	−23
Race (62)				
White	58	13	4	− 5
Nonwhite	239	25	11	−43
Residence (64)				
Rural nonfarm	52	18	− 2	−23
Rural farm	37	19	a	6
SMSA central city	153	39	−11	−20
SMSA other urban	37	11	33	10
Urban nonSMSA	71	23	−20	− 7
Level of use				
Expenditures per admission (23)[b]				
Low ($0 to 300)	22	4	1	3
Medium ($301 to 600)	45	9	3	1
High ($601 or more)	173	42	10	−35
Total	$ 76	$13	$ 5	$ − 9

[a]Less than fifty cents.

[b]Admissions discovered in verification are omitted.

Table 4.15. Differences between Groups by Unadjusted and Adjusted Social Survey Estimates for Out-of-Pocket and Insurance Payments per Hospital Admission

Characteristic	*Out-of-Pocket Payments per Admission (53)*			*Insurance Payments per Admission (51)*		
	Unadjusted Difference	*Adjusted Difference*	*Standard Error of the Difference*	*Unadjusted Difference*	*Adjusted Difference*	*Standard Error of the Difference*
Demographic						
Age of oldest family member (7)						
Less than 65 versus						
65 or more	$ 67**	$ 29	$29.8	$ 262**	$ 272**	$ 42.9
Family income (26)						
Nonpoor versus poor	30	35*	34.5	232**	252**	52.2
Race (62)						
White versus nonwhite	77**	60**	27.0	89	83	92.1
Residence (64)						
SMSA other urban versus:						
Rural nonfarm	114*	108*	75.8	181*	192*	125.0
Rural farm	98*	79	80.3	268**	301**	112.3
SMSA central city	125*	115*	75.9	145*	130*	118.5
Urban nonSMSA	119*	112*	75.4	277**	284**	113.5
Level of use						
Expenditures per admission (23)[a]						
Low ($0 to 300) versus:						
Medium ($301 to 600)	− 57**	− 40**	8.9	−115**	−124**	14.3
High ($601 or more)	−208**	−138*	77.0	−603**	−588**	86.3

Note: * indicates that the difference shown is greater than one standard error; ** indicates that the difference shown is greater than two standard errors.

[a]Admissions discovered in verification are omitted.

Table 4.16. Differences between Groups by Unadjusted and Adjusted Social Survey Estimates for Medicare and Welfare Payments per Hospital Admission

	Medicare Payments per Admission (52)			Welfare Payments per Admission (54)		
Characteristic	*Unadjusted Difference*	*Adjusted Difference*	*Standard Error of the Difference*	*Unadjusted Difference*	*Adjusted Difference*	*Standard Error of the Difference*
Demographic						
Age of oldest family member (7)						
Less than 65 versus 65 or more	$ —[a]	$ —[a]	$ —[a]	$ 24	$ 48**	$23.8
Family income (26)						
Nonpoor versus poor	−120**	−140**	37.4	−190**	−140**	41.1
Race (62)						
White versus nonwhite	24	7	25.4	−181**	−150**	28.1
Residence (64)						
SMSA other urban versus:						
Rural nonfarm	18	21	44.6	− 16	52**	20.9
Rural farm	− 64*	− 62*	55.5	− 1	37*	21.8
SMSA central city	7	− 23	44.4	−117**	− 42*	40.9
Urban nonSMSA	48*	30	44.7	− 34	37*	25.8
Level of use						
Expenditures per admission (23)[b]						
Low ($0 to 300) versus:						
Medium ($301 to 600)	− 37**	− 37**	7.5	− 23**	− 22**	10.0
High ($601 or more)	−286**	−276**	41.7	−151**	−122**	41.8

Note: * indicates that the difference shown is greater than one standard error; ** indicates that the difference shown is greater than two standard errors.

[a]Not applicable.

[b]Admissions discovered in verification are omitted.

Table 5.1. Distribution of Social Survey and Verification Physician Visit Conditions by Degree of Match

Characteristic	Social Survey Conditions				Verification Diagnoses				Summary Accuracy Score
	Percent First Degree Match	Percent Second Degree Match	Percent Un-matched	Total Percent	Percent First Degree Match	Percent Second Degree Match	Percent Un-matched	Total Percent	
Demographic									
Age (6)[a]									
Less than 6 years	41%	20%	39%	100%	37%	14%	49%	100%	56% (2.7)[b]
6 to 17	42	15	43	100	45	16	40	100	59 (2.0)
18 to 34	42	20	38	100	37	18	45	100	58 (1.7)
35 to 54	39	14	47	100	33	18	49	100	52 (2.1)
55 to 64	40	19	41	100	31	18	51	100	53 (1.9)
65 years or more	37	23	40	100	29	24	48	100	55 (1.5)
Sex (68)									
Male	42	18	40	100	37	18	45	100	57 (1.1)
Female	39	18	42	100	35	17	49	100	55 (.9)
Race (62)									
White	41	18	41	100	36	18	46	100	56 (.8)
Nonwhite	37	15	48	100	33	17	51	100	51 (2.4)
Family income (26)									
Nonpoor	41	18	41	100	36	18	46	100	56 (.8)
Poor	38	19	42	100	33	18	49	100	54 (1.2)
Residence (64)									
Rural nonfarm	39	17	44	100	34	17	49	100	53 (1.4)
Rural farm	38	13	49	100	36	17	48	100	51 (2.8)
SMSA central city	43	18	39	100	35	18	47	100	57 (1.4)
SMSA other urban	40	20	40	100	36	18	46	100	57 (1.7)
Urban nonSMSA	41	18	41	100	38	19	43	100	58 (2.3)

Table 5.1. (Continued)

Characteristic	Social Survey Conditions				Verification Diagnoses				Summary Accuracy Score
	Percent First Degree Match	*Percent Second Degree Match*	*Percent Un-matched*	*Total Percent*	*Percent First Degree Match*	*Percent Second Degree Match*	*Percent Un-matched*	*Total Percent*	
Survey respondent (71)									
Self	41	19	41	100	33	19	48	100	55 (1.0)
Proxy for child 18 years or less	42	17	41	100	41	15	44	100	58 (1.6)
Proxy for adult	38	18	44	100	33	19	48	100	54 (1.6)
Threat									
Threat of physician visit conditions[c] (72)									
Very threatening	51	12	36	100	29	10	61	100	46 (1.9)
Moderately threatening	46	21	33	100	28	15	57	100	52 (1.3)
Nonthreatening	40	18	42	100	38	19	43	100	57 (.8)
Perceived and evaluated health									
Perception of health[c] (55)									
Excellent	42	16	43	100	41	15	44	100	57 (1.4)
Good	41	18	41	100	35	18	47	100	56 (1.2)
Fair	37	19	44	100	32	20	48	100	54 (1.4)
Poor	38	23	40	100	28	25	47	100	56 (2.6)
Worry about health[c] (73)									
A great deal	42	24	34	100	32	23	45	100	60 (1.5)
Some	39	18	43	100	33	20	47	100	55 (1.5)
Hardly any	37	18	45	100	34	17	50	100	53 (1.4)
None	43	14	43	100	40	15	45	100	56 (1.5)
Impact of illness									
Severity of physician visit condition (66)									
Preventive care	30	14	56	100	46	18	36	100	52 (1.4)

Symptomatic relief available	40	22	37	100	34	16	50	100	55 (1.2)
Should see doctor	47	23	30	100	30	19	51	100	57 (1.3)
Must see doctor	57	14	30	100	39	17	44	100	62 (1.5)
Physician visit conditions per person[d] (47)									
1 to 2	43	17	41	100	35	17	48	100	55 (1.3)
3 to 5	36	21	44	100	41	24	35	100	60 (1.3)
6 or more	42	26	32	100	46	33	21	100	73 (6.0)
Physician visits per person (59)									
1	35	12	54	100	36	14	50	100	48 (2.1)
2 to 4	40	16	44	100	36	15	49	100	53 (1.2)
5 to 13	43	20	37	100	36	19	45	100	59 (1.2)
14 or more	45	22	33	100	36	23	41	100	63 (1.7)
Out-of-pocket expenditures for physician visits (22)									
$0 to 15	38	15	47	100	35	16	49	100	52 (1.7)
16 to 45	40	18	42	100	35	17	48	100	55 (1.6)
46 or more	45	21	35	100	36	20	43	100	61 (.7)
Expenditures for prescribed drugs (25)									
$0 to 25	40	17	43	100	37	16	48	100	55 (1.0)
26 to 75	41	18	41	100	34	18	47	100	55 (1.8)
76 or more	42	23	35	100	32	25	42	100	61 (1.5)
Total	40%	18%	42%	100%	36%	18%	47%	100%	56% (.7)

[a]In this and all subsequent tables in the chapter, the numbers in parentheses refer to the variable definition list found in Appendix C.

[b]The numbers in parentheses in this column are the standard errors of the accuracy scores. Significant differences between subgroups, when discussed in the text, refer to differences at least twice the standard error (SE) of the difference. The SE_{diff} is computed as the square root of the sum of the squares of the two individual accuracy score standard errors.

[c]"Don't know" and "no answer" responses are excluded.

[d]Cases where individual reported none in social survey are excluded.

Table 5.2. Error Components in Social Survey Estimates of Physician Visit Condition Rates per 1,000 Persons

Physician Visit Condition[a]	*Social Survey Rate per 1,000 Persons*	*Field Bias*	*Standard Error*
Checkup	146.85	−70.50	3.28
Tests and shots	85.74	−45.75	2.60
Common cold and other acute upper respiratory conditions	65.51	− 1.88	2.30
Visual exams and impairments	54.72	−10.30	2.11
Pregnancy/delivery	24.38	− .47	1.43
High blood pressure	23.35	3.93	1.40
Influenza (nondigestive manifestations)	19.08	− 4.89	1.27
Sprains, strains, whiplash, torn ligaments	17.79	.29	1.23
Diabetes	15.75	− .78	1.16
Heart trouble	15.70	4.34	1.15
Other acute musculo-skeletal diseases	15.05	3.22	1.13
Allergies	15.03	5.54	1.13
Minor and intermediate psychological problems	14.80	14.72	1.12
"Female trouble"—chronic	13.41	10.02	1.07
Acute diseases of the ear	12.64	8.48	1.04
Open wounds and lacerations	12.41	2.99	1.03
Arthritis and rheumatism	12.41	4.24	1.03
Benign tumor	11.07	6.38	.97
Contusions and superficial injuries	9.80	5.09	.91
Fractures and dislocations	9.61	.70	.90
Chronic diseases of the urinary system and male genital system	9.34	3.83	.89
Other acute genitourinary disorders	9.24	5.46	.89

Note: Conditions with an unweighted social survey N of less than 100 are excluded.

[a]For provider-reported conditions, subcategories of the 99 conditions codes have been collapsed to permit rate and bias estimation for specific conditions.

Table 5.3. Percentage Distribution of Social Survey and Verification Hospitalized Conditions by Degree of Match

Characteristic	Social Survey Conditions				Verification Diagnoses				Summary Accuracy Score
	Percent First Degree Match	Percent Second Degree Match	Percent Un-matched	Total Percent	Percent First Degree Match	Percent Second Degree Match	Percent Un-matched	Total Percent	
Demographic									
Age (6)									
Less than 6 years	64%	7%	28%	100%	47%	7%	46%	100%	61% (5.0)[a]
6 to 17	58	17	25	100	52	14	33	100	70 (4.9)
18 to 34	45	21	34	100	41	22	31	100	68 (4.5)
35 to 54	64	11	25	100	50	20	31	100	72 (3.2)
55 to 64	64	18	18	100	34	22	44	100	65 (4.5)
65 years or more	50	21	29	100	32	24	45	100	61 (2.4)
Sex (68)									
Male	58	16	26	100	42	20	38	100	67 (2.6)
Female	54	17	28	100	42	20	38	100	66 (2.6)
Race (62)									
White	58	17	25	100	43	20	37	100	68 (1.9)
Nonwhite	41	16	43	100	35	19	45	100	56 (4.7)
Family income (26)									
Nonpoor	57	17	27	100	44	20	36	100	68 (2.6)
Poor	54	17	29	100	37	19	45	100	61 (2.4)
Residence (64)									
Rural nonfarm	59	17	24	100	43	20	37	100	68 (2.6)
Rural farm	57	17	27	100	43	25	33	100	70 (3.1)
SMSA central city	52	14	34	100	38	16	47	100	59 (3.2)
SMSA other urban	56	18	26	100	44	20	35	100	69 (4.2)
Urban nonSMSA	56	21	23	100	43	24	33	100	71 (2.2)

Table 5.3. (Continued)

Characteristic	*Social Survey Conditions*				*Verification Diagnoses*				*Summary Accuracy Score*	
	Percent First Degree Match	*Percent Second Degree Match*	*Percent Un-matched*	*Total Percent*	*Percent First Degree Match*	*Percent Second Degree Match*	*Percent Un-matched*	*Total Percent*		
Survey respondent (71)										
Self	56	19	25	100	41	22	37	100	68	(2.2)
Proxy for child										
18 years or less	59	13	28	100	50	12	38	100	66	(3.6)
Proxy for adult	53	15	32	100	39	20	41	100	63	(3.2)
Threat										
Threat of physician visit conditions[b] (72)										
Very threatening	70	7	23	100	62	6	32	100	72	(3.6)
Moderately threatening	51	24	25	100	36	23	41	100	65	(3.1)
Nonthreatening	54	17	29	100	39	22	39	100	65	(2.3)
Perceived and evaluated health										
Perception of health[b] (55)										
Excellent	57	18	25	100	50	16	33	100	71	(4.8)
Good	61	14	25	100	49	18	33	100	70	(3.0)
Fair	50	21	28	100	35	24	40	100	64	(2.8)
Poor	46	21	33	100	34	27	40	100	63	(3.5)
Worry about health (73)										
A great deal	51	18	31	100	36	21	43	100	62	(2.7)
Some	58	15	27	100	45	18	37	100	67	(2.8)
Hardly any	54	17	29	100	47	22	31	100	70	(5.2)
None	65	20	15	100	49	22	28	100	77	(3.0)
Number of diagnoses[c] (45)										
1	55	15	30	100	47	17	36	100	67	(3.9)
2	61	17	22	100	43	23	34	100	71	(2.8)
3	53	20	28	100	43	25	32	100	70	(3.1)
4 or more	56	17	27	100	39	18	42	100	64	(2.8)

Impact of illness									
Number of hospitalized conditions[c] (46)									
1	68	15	17	100	45	20	36	100	71 (2.1)
2 to 3	41	19	41	100	41	23	36	100	62 (3.2)
4 or more	42	35	23	100	37	35	28	100	74 (4.3)
Number of hospital days per condition[c] (34)									
1 to 3	55	14	32	100	54	18	27	100	71 (3.9)
4 to 7	56	21	24	100	46	23	31	100	72 (2.9)
8 or more	57	16	27	100	40	20	40	100	65 (2.1)
Operations per hospitalized condition (50)									
0	46	21	33	100	32	24	45	100	60 (1.6)
1 or more	67	13	20	100	57	14	29	100	75 (2.9)
Hospital admissions per condition[c] (29)									
1	55	17	29	100	46	21	33	100	69 (1.9)
2 or more	63	19	19	100	36	20	45	100	64 (3.9)
Out-of-pocket hospital expenditures per condition (21)									
$0	53	19	27	100	40	20	39	100	65 (2.0)
1 to 200	58	13	29	100	45	20	35	100	67 (3.0)
201 or more	79	12	9	100	45	16	39	100	70 (9.9)
Severity of hospitalized condition[b,c] (66)									
Preventive care	10	11	80	100	23	21	56	100	27 (6.4)
Symptomatic relief available	42	24	34	100	33	29	37	100	63 (3.6)
Should see doctor	62	19	19	100	41	22	38	100	70 (2.5)
Must see doctor	59	12	29	100	47	15	38	100	66 (2.1)
Total	56%	17%	27%	100%	42%	20%	38%	100%	66% (1.8)

[a]The numbers in parentheses in this column are the standard errors of the accuracy scores. See Table 5.1, footnote b.

[b] "Don't know" and "no answer" responses are excluded.

[c]Cases where individual reported none in social survey are excluded.

Table 5.4. Error Components in Social Survey Estimates of Hospitalized Condition Rates per 1,000 Persons

Hospitalized Condition[a]	*Social Survey Rate per 1,000 Persons*	*Field Bias*	*Standard Error*
Pregnancy/delivery	23.21	.04	1.40
Chronic diseases of urinary system and male genital system	6.79	.71	.76
Tonsilitis and adenoiditis	5.21	.19	.67
Fractures and dislocations	5.00	− .84	.65
Sprains, strains, whiplash, torn ligaments	4.98	−1.39	.65
"Female trouble"—chronic	4.84	.33	.64
Heart trouble	4.22	2.81	.60
Other acute musculo-skeletal diseases	3.91	.43	.58
Benign tumor	3.70	.37	.56
Diabetes	3.23	2.05	.53
Diseases of the gall bladder	3.20	.32	.52
Hernia	2.76	.26	.49
Cancer	2.66	.14	.48
Hospitalized mental illness	2.59	1.59	.47
Complications of procedures or poisoning	2.50	.41	.46
Unspecified tumor	2.31	− .51	.44
Other visual impairments	2.13	.72	.43
Other serious chronic diseases of the digestive system	1.97	1.23	.41

Note: Conditions with an unweighted social survey N of less than 30 are excluded.

[a]For provider-reported conditions, subcategories of the 99 conditions codes have been collapsed to permit rate and bias estimation for specific conditions.

Table 5.5. Distribution of Social Survey and Verification Surgical Procedures by Degree of Match

Characteristic	*Social Survey Surgical Procedures*				*Verification Surgical Procedures*				*Summary Accuracy Score*
	Percent First Degree Match	*Percent Second Degree Match*	*Percent Un-matched*	*Total Percent*	*Percent First Degree Match*	*Percent Second Degree Match*	*Percent Un-matched*	*Total Percent*	
Demographic									
Age (6)									
Less than 6 years	85%	10%	6%	100%	82%	9%	8%	100%	93% (2.6)[a]
6 to 17	80	6	14	100	74	5	20	100	82 (5.2)
18 to 34	70	16	14	100	64	14	22	100	82 (5.3)
35 to 54	80	4	16	100	80	4	15	100	85 (4.0)
55 to 64	85	3	12	100	70	3	27	100	80 (4.6)
65 years or more	62	12	26	100	64	12	25	100	75 (5.1)
Sex (68)									
Male	79	8	13	100	77	7	16	100	86 (2.4)
Female	74	10	17	100	68	9	23	100	80 (3.1)
Race (62)									
White	79	9	13	100	73	8	19	100	84 (2.4)
Nonwhite	51	9	40	100	54	9	38	100	61 (8.2)
Family income (26)									
Nonpoor	78	9	13	100	72	9	19	100	84 (2.4)
Poor	69	6	25	100	70	6	24	100	76 (3.8)
Residence (64)									
Rural nonfarm	72	12	16	100	67	11	22	100	81 (3.7)
Rural farm	81	4	15	100	74	3	23	100	81 (6.3)
SMSA central city	70	4	26	100	65	4	31	100	71 (6.3)
SMSA other urban	77	14	8	100	76	14	11	100	90 (2.1)
Urban nonSMSA	89	5	5	100	81	5	14	100	90 (2.7)

Table 5.5. (Continued)

Characteristic	Social Survey Surgical Procedures				Verification Surgical Procedures				Summary Accuracy Score
	Percent First Degree Match	*Percent Second Degree Match*	*Percent Un-matched*	*Total Percent*	*Percent First Degree Match*	*Percent Second Degree Match*	*Percent Un-matched*	*Total Percent*	
Survey respondent (71)									
Self	77	8	15	100	73	7	20	100	82 (2.9)
Proxy for child 18 years or less	82	7	11	100	77	7	16	100	86 (4.3)
Proxy for adult	67	14	19	100	61	13	26	100	77 (4.8)
Threat									
Threat of operation (72)									
Nonthreatening	79	8	13	100	71	8	21	100	81 (3.5)
Threatening	75	9	15	100	72	8	20	100	85 (2.8)
Perceived and evaluated health									
Perception of health[b] (55)									
Excellent	79	9	12	100	77	9	14	100	87 (3.4)
Good	80	9	11	100	72	8	20	100	84 (3.5)
Fair	69	9	21	100	68	9	23	100	78 (5.6)
Poor	68	8	23	100	69	7	24	100	76 (5.7)
Worry about health[b] (73)									
A great deal	70	11	19	100	70	11	18	100	82 (5.3)
Some	74	10	16	100	70	9	21	100	82 (3.6)
Hardly any	85	2	13	100	72	2	26	100	80 (6.0)
None	83	8	8	100	77	7	16	100	88 (3.0)
Number of diagnoses[c] (45)									
1	84	4	12	100	74	4	23	100	82 (5.0)
2	74	8	18	100	71	8	22	100	80 (3.8)

3	73	10	17	100	74	11	16	100	84 (3.9)
4 or more	75	11	14	100	72	10	18	100	84 (4.1)
Impact of illness									
Number of social survey surgical procedures[c] (48)									
1	79	8	14	100	82	8	10	100	88 (2.0)
2 or more	63	14	23	100	70	15	15	100	81 (8.5)
Hospital days for stay of operation (32)									
0 to 3	79	8	13	100	81	9	11	100	88 (2.4)
4 to 5	76	6	18	100	76	6	18	100	82 (5.9)
6 or more	74	10	16	100	71	9	20	100	82 (3.2)
Hospital expenditures for stay of operation (20)									
$1 to 300	75	10	15	100	75	10	16	100	85 (2.6)
301 to 600	79	5	16	100	79	15	16	100	84 (3.1)
601 or more	74	11	15	100	71	11	18	100	84 (3.7)
Total	76%	8%	15%	100%	72%	8%	20%	100%	82% (2.3)

[a]The numbers in parentheses in this column are the standard errors of the accuracy scores. See Table 5.1, footnote b.

[b]"Don't know" and "no answer" responses are excluded.

[c]Cases where individual reported none in social survey are excluded.

Table 5.6. Error Components in Social Survey Estimates of Surgical Procedures Rates per 1,000 Persons

Surgical Procedure	*Social Survey Rate per 1,000 Persons*	*Field Bias*	*Standard Error*
Normal delivery	16.14	− .03	1.17
Setting of fracture, dislocation	6.69	−1.57	.76
Tonsillectomy, adenoidectomy	5.00	.03	.65
Operations on hernia	2.83	− .01	.49
Cholecystectomy	2.61	.00	.47
Delivery with complications	2.34	− .02	.45
Operation on bladder	2.29	.34	.44
D&C for pregnancy	2.27	.40	.44
D&C (non-pregnancy-related)	2.17	− .22	.43
Hysterectomy	2.08	− .03	.42
Hemorrhoidectomy, fistulectomy	1.87	.28	.40
Other operations on digestive system (liver operation, intussusception)	1.83	.06	.40
Appendectomy	1.44	− .11	.35
Cataract operation	1.23	.00	.33
Other operations on female genital organs (tubes tied, hymenotomy)	1.09	.63	.31
Breast surgery	.97	.12	.29
Operation on skin	.97	.52	.29

Note: Procedures with an unweighted social survey N of less than 20 are excluded.

Table 5.7. Factors Significantly Associated with Accuracy of Reporting

	Physician Visit Conditions	*Hospitalized Conditions*	*Surgical Procedures*
Demographic characteristics	SEX (female –)[a]	—	AGE (5 years or less +)
	RACE (nonwhite –)	RACE (nonwhite –)	RACE (nonwhite –)
	RESIDENCE (rural farm –)	RESIDENCE (central city –)	RESIDENCE (central city – , other urban +)
	—	FAMILY INCOME (poor –)	FAMILY INCOME (poor –)
Threat	THREAT (very threatening –)	—	—
Perceived and evaluated health	WORRY (great deal +)	WORRY (none +)	—
Impact of condition	NUMBER OF CONDITIONS (1 or 2 – , 6 or more +)	NUMBER OF CONDITIONS (2 or 3 –)	—
	SEVERITY (must see MD +)	SEVERITY (preventive only –)	—
	PHYSICIAN VISITS (one – , 5 or more +)	HOSPITAL DAYS (8 or more –)	—
	OUT-OF-POCKET EXPENSES ($46 or more +)	OPERATION DURING STAY (one or more +)	—
	DRUG EXPENSES ($76 or more +)	—	—

Note: Differences between subgroups for a given factor are greater than two standard errors of the difference.

[a]The terms in parentheses indicate the varying category and its level of accuracy (+ = higher, – = lower than the other categories for the factor).

Table 6.1. Response to Verification Request by Characteristics of Respondents

Characteristic	Response to Verification Request (65)		Standard Error
	Person Gave Permisssion to Verify	Person Refused Permission to Verify	
Demographic			
Age of oldest family member (7)[a]			
Less than 65 years	87.7%	12.3%	1.2
65 years or more	88.2	11.8	1.5
Family income (26)			
Nonpoor	86.0	4.0	1.2
Poor	93.8	6.2	.6**
Race (62)			
White	87.6	12.4	1.0
Nonwhite	89.1	10.9	2.8
Residence (64)			
Rural nonfarm	92.2	7.8	1.7**
Rural farm	92.8	7.2	2.2**
SMSA central city	85.7	14.3	1.7
SMSA other urban[b]	81.9	18.1	3.0
Urban nonSMSA	94.3	5.7	1.0**
Cooperation			
Income was reported (37)			
Yes	88.8	11.2	.9
No	63.6	36.4	5.8**
Hospital expenditures were reported (36)			
Yes	93.0	7.0	1.2
No	92.9	7.1	2.0
Physician expenditures were reported (58)			
Yes	89.8	10.2	2.0
No	87.6	12.4	1.2
Cooperation on calls by interviewer (11)			
Cooperative	89.6	10.4	1.0
Uncooperative	72.5	27.5	5.2**
Number of calls to complete main questionnaire (44)			
1[b]	91.9	8.1	1.6
2	92.4	7.6	1.6
3	85.4	14.6	2.5**
4	84.8	15.2	3.9
5 or more	77.1	22.9	4.2**

Table 6.1. (Continued)

Characteristic	*Response to Verification Request (65)*		*Standard Error*
	Person Gave Permisssion to Verify	*Person Refused Permission to Verify*	
Perceived and evaluated health			
Perception of health (55)			
Excellent[b]	87.0	13.0	1.9
Good	87.0	13.0	1.1
Fair	90.2	9.8	1.1
Poor	92.4	7.6	2.1
Worry about health (73)			
A great deal	92.0	8.0	1.5**
Some	90.3	9.7	1.4**
Hardly any	87.1	12.9	1.5
None[b]	86.3	13.7	1.4
Severity over all diagnoses (67)			
Elective only[b]	80.7	19.3	1.8
Elective and mandatory	91.5	8.5	1.2**
Mandatory only	91.2	8.8	1.0**
Number of diagnoses (45)			
0[b]	88.8	11.2	1.2
1	84.5	15.5	1.2**
2	84.6	15.4	2.1
3	91.1	8.9	1.3
4 or more	96.5	3.5	1.3**
Total	87.8%	12.2%	1.0

Note: ** indicates that the difference between the compared categories is greater than two standard errors.

[a]In this and all subsequent tables in the chapter, the numbers in parentheses refer to the variable definition list found in Appendix C.

[b]Category against which others were compared to test the statistical significance of the difference.

Table 6.2. Cooperation with Verification by Characteristics of Physicians

Characteristic	*Physician Cooperation*[a] *(57)*		*Standard Error*
	Cooperated	*Did Not Cooperate*	
Board certified (9)			
Yes	90.3%	9.7%	1.2
No	88.5	11.5	.8
Age of physician (8)			
Less than 36 years	87.3	12.7	2.5
36 to 45	88.3	11.7	1.5
46 to 55	89.0	11.0	1.3
56 to 65	90.9	9.1	1.2
66 years or more[b]	86.6	13.4	3.0
Specialty (70)			
General practice[b]	88.4	11.6	1.0
General surgery	86.2	13.8	2.9
Primary care specialties[c]	88.8	11.2	1.5
Other specialties	91.2	8.8	1.0**

Note: These characteristics are available only for physicians listed in the 1969 AMA Directory.

** indicates that the difference between the compared categories is greater than two standard errors.

[a]A physician appears more than once in this table if he or she was asked to verify survey data for more than one respondent. Physicians are assigned the weights of respondents and may appear both as having cooperated and as not having cooperated, depending on their response to verification for each patient.

[b]Category against which others were compared to test the significance of the standard error of the difference.

[c]Includes internal medicine, obstetrics-gynecology, and pediatrics.

Table 6.3. Cooperation with Verification by the Number of Times Verification Was Requested

Number of Times Verification Requested (49)	*Physicians*			*Clinics*		
	Always Cooperated	*Cooperated Sometimes*	*Did Not Cooperate*	*Always Cooperated*	*Cooperated Sometimes*	*Did Not Cooperate*
1	91.3%	—%	8.7%	93.3%	—%	6.7%
2	84.3	4.4	11.3	88.2	3.2	8.6
3	85.8	5.2	9.0	83.9	14.1	2.0
4 to 5	87.5	5.1	6.6	73.5	9.2	17.2
6 to 10	85.4	7.4	7.1	78.9	.9	20.1
11 to 20	67.4	20.9	11.6	52.8	38.2	9.0
21 or more	51.7	40.4	7.8	56.4	43.6	—
Total	83.3%	6.6%	10.1%	78.4%	12.3%	9.3%

Note: The weight for each physician and clinic is the sum of the weights of all persons naming them.

Table 6.4. Reporting Accuracy for Hospital Admissions, Days, and Expenditures by Characteristics of Respondents

Characteristic	Reporting Accuracy For Admissions (1)			Reporting Accuracy For Days (2)			Reporting Accuracy For Expenditures (3)		
	Under-reported	*Over-reported*	*Accurately Reported*	*Under-reported*	*Over-reported*	*Accurately Reported*	*Under-reported*	*Over-reported*	*Accurately Reported*
Demographic									
Age (6)									
Less than 17 years	8.3%	6.6%	85.1%	9.5%	16.0%	74.5%	14.2%	6.2%	79.6%
17 to 40	4.7	5.4	89.9	4.7	10.1	85.2	6.5	11.2	82.3
41 to 64	2.0	7.0	91.0	4.2	10.3	85.5	5.4	7.8	86.8
65 years or more	5.5	12.0	82.5	4.9	19.2	75.9	15.0	8.9	76.1
Family income (26)									
Nonpoor	3.9	5.2	91.0	5.2	10.2	84.6	6.0	8.1	85.9
Poor	7.5	12.4	80.1	6.4	19.8	73.9	20.0	14.0	66.0
Race (62)									
White	4.7	6.5	88.8	5.4	11.7	83.0	7.7	9.0	83.3
Nonwhite	5.5	11.5	83.1	6.6	21.2	72.2	20.0	11.7	67.6
Residence (64)									
Rural nonfarm	6.1	5.5	88.4	7.1	12.7	80.2	9.8	7.9	82.4
Rural farm	3.5	9.1	87.4	3.8	12.7	83.5	8.5	14.1	77.4
SMSA central city	6.2	9.9	83.8	8.6	12.9	78.4	8.7	13.7	77.6
SMSA other urban	1.7	3.7	94.6	2.6	10.3	87.1	5.0	5.4	89.6
Urban nonSMSA	6.1	9.1	84.8	3.0	15.9	81.0	13.7	9.0	77.2
Education (12)									
8 years or less	4.2	12.0	83.9	5.9	15.2	78.9	11.5	9.4	79.1
9 to 11	8.6	6.2	85.2	9.0	14.2	76.8	13.4	11.6	75.0
12	2.0	6.5	91.5	3.0	9.0	88.0	5.4	12.5	82.1
13 years or more	2.1	3.4	94.5	.7	9.0	90.4	2.8	5.0	92.2

Survey Respondent (71)									
Self	3.8	6.9	89.2	4.5	10.8	84.7	7.2	9.1	83.7
Proxy for child 16 years or less	7.9	6.6	85.4	9.6	16.1	74.3	14.0	6.4	79.6
Proxy for adult	4.8	7.4	87.7	4.9	14.2	80.9	8.4	11.3	80.3
Perceived and evaluated health									
Perception of health (55)									
Excellent	3.9	3.8	92.3	4.2	7.9	87.9	10.1	7.1	82.8
Good	4.9	6.2	88.9	6.5	12.7	80.8	7.0	9.1	83.9
Fair	4.4	10.3	85.3	3.7	14.7	81.6	11.8	11.2	77.0
Poor	4.4	11.2	84.5	4.2	19.1	76.8	6.9	10.8	82.3
Worry about health (73)									
A great deal	4.6	7.5	87.8	5.5	15.2	79.3	9.0	7.1	83.8
Some	4.3	6.6	89.1	5.0	10.9	84.2	7.7	10.5	81.8
Hardly any	6.6	11.3	82.1	5.1	20.4	74.5	6.3	14.0	79.7
None	3.6	3.9	92.5	5.4	6.4	88.2	11.3	5.3	83.4
Severity over all diagnoses (67)									
Elective only	6.2	6.2	87.6	7.6	6.6	85.8	1.9	9.4	88.7
Elective and mandatory	2.4	7.5	90.1	3.3	12.6	84.1	7.7	8.7	83.6
Mandatory only	6.0	5.6	88.4	6.4	11.8	81.8	9.5	8.8	81.6
Number of diagnoses (45)									
1	5.6	6.8	87.6	6.0	12.3	81.7	6.1	11.8	82.1
2	3.3	6.1	90.6	3.9	14.8	81.2	13.0	9.8	77.1
3	4.4	9.8	85.8	4.5	14.3	81.2	3.7	10.6	85.7
4 or more	3.0	4.7	92.3	4.2	8.1	87.7	8.0	4.4	87.5
Cooperation									
Cooperation on calls by interviewer (11)									
Cooperative	4.6	7.0	88.4	5.5	12.2	82.3	8.0	8.5	83.5
Uncooperative	6.7	7.4	85.9	5.1	17.3	77.6	14.8	15.6	69.6

Table 6.4. (Continued)

Characteristic	Reporting Accuracy For Admissions (1)			Reporting Accuracy For Days (2)			Reporting Accuracy For Expenditures (3)		
	Under-reported	*Over-reported*	*Accurately Reported*	*Under-reported*	*Over-reported*	*Accurately Reported*	*Under-reported*	*Over-reported*	*Accurately Reported*
Number of calls to complete main questionnaire (44)									
1	4.2	5.5	90.3	5.8	13.3	80.9	7.6	6.6	85.8
2	4.6	8.3	87.1	3.8	11.8	84.4	7.2	6.6	86.2
3	3.7	5.8	90.5	3.3	9.1	87.5	11.6	7.1	81.3
4	6.4	3.4	90.1	8.1	10.1	81.8	9.7	7.1	83.2
5 or more	6.7	12.6	80.6	8.1	18.2	73.7	9.4	24.0	66.5
Level of use									
Hospital days per person with admission (35)									
1 to 4	1.3	7.6	91.1	2.3	13.3	84.4	7.2	9.2	83.6
5 to 10	2.2	6.5	91.4	2.8	11.5	85.7	4.5	11.2	84.4
11 or more	3.2	7.4	89.4	3.5	14.3	82.2	8.6	7.6	83.8
Expenditures per person with admission (19)									
$0 to 300	3.6	9.7	86.7	3.8	18.2	78.1	14.1	11.6	74.3
301 to 600	.9	6.6	92.4	1.9	10.8	87.3	2.7	5.9	91.4
601 or more	2.2	5.7	92.1	2.9	10.9	86.2	5.1	11.2	83.7
Total	4.8%	7.0%	88.2%	5.5%	12.6%	81.9%	8.6%	9.2%	82.2%

Note: In this table and all subsequent tables in the chapter, the characteristics shown are those of the designated respondent rather than the person who may have responded for him or her.

Table 6.5. Reporting Accuracy for Physician Visit and Office Visit Expenditures by Characteristics of Respondents

Characteristic	*Reporting Accuracy for Physician Visits (4)*			*Reporting Acuuracy for Office Visit Expenditures (5)*		
	Under-reported	*Over-reported*	*Accurately Reported*	*Under-reported*	*Over-reported*	*Accurately Reported*
Demographic						
Age (6)						
Less than 17 years	19.6%	27.9%	52.5%	23.5%	31.8%	44.7%
17 to 40	18.7	27.7	53.5	24.4	30.9	44.6
41 to 64	17.8	27.9	54.2	22.1	34.7	43.2
65 years or more	22.8	24.4	52.7	23.8	27.1	49.1
Family income (26)						
Nonpoor	19.2	25.8	55.0	22.5	32.0	45.5
Poor	19.6	34.0	46.4	28.0	30.3	41.7
Race (62)						
White	19.7	26.3	54.1	23.5	31.2	45.3
Nonwhite	15.4	38.6	46.0	22.7	38.7	38.6
Residence (64)						
Rural nonfarm	19.3	26.5	54.2	23.1	32.9	44.0
Rural farm	22.0	26.7	51.3	24.6	29.3	46.1
SMSA central city	19.4	30.7	49.8	25.1	29.0	45.8
SMSA other urban	20.3	25.1	54.6	23.5	33.7	42.7
Urban nonSMSA	15.5	27.6	57.0	20.6	31.5	47.9
Education (12)						
Less than 9 years	17.5	31.1	51.4	23.2	33.7	43.1
9 to 11	19.6	27.0	53.4	26.0	28.7	45.4
12	20.5	27.3	52.1	24.7	33.4	41.8
13 years or more	20.2	22.7	57.1	21.7	29.3	49.1

Table 6.5. (Continued)

Characteristic	Reporting Accuracy for Physician Visits (4)			Reporting Acuuracy for Office Visit Expenditures (5)		
	Under-reported	*Over-reported*	*Accurately Reported*	*Under-reported*	*Over-reported*	*Accurately Reported*
Survey Respondent (71)						
Self	18.3	26.8	55.0	22.5	31.0	46.5
Proxy for child 16 years or less	19.6	27.9	52.5	23.5	31.7	44.9
Proxy for adult	20.7	28.1	51.2	25.3	33.1	41.7
Perceived and evaluated health						
Perception of health (55)						
Excellent	20.3	21.4	55.2	26.8	27.3	45.9
Good	20.1	27.6	52.3	25.1	31.0	44.0
Fair	12.4	34.7	53.0	16.9	36.8	46.3
Poor	11.8	40.1	48.1	12.5	48.5	39.0
Worry about health (73)						
A great deal	13.5	38.7	47.8	18.2	42.2	39.6
Some	13.3	31.1	55.6	20.1	33.6	46.3
Hardly any	19.9	26.5	53.6	25.2	28.8	45.9
None	25.5	21.6	52.9	26.9	28.9	44.2
Severity over all diagnoses (67)						
Elective only	19.6	26.0	54.3	23.0	30.6	46.5
Elective and mandatory	20.6	25.9	53.5	23.3	30.6	46.1
Mandatory only	17.3	30.6	52.1	24.1	34.4	41.5
Number of diagnoses (45)						
1	16.3	32.0	51.7	21.4	34.8	43.8
2	22.5	24.4	53.1	25.9	33.0	41.1
3	20.3	24.5	55.3	24.4	25.8	49.8
4 or more	19.7	25.2	55.2	23.1	28.7	48.1

Cooperation						
Cooperation on calls by interviewer (11)						
Cooperative	19.5	27.3	53.2	24.2	31.4	44.4
Uncooperative	16.7	29.1	54.1	16.8	34.6	48.7
Number of calls to complete main questionnaire (44)						
1	17.6	29.2	53.2	23.4	31.8	44.8
2	21.3	25.0	53.8	24.3	31.4	44.3
3	21.0	25.1	53.9	23.9	32.5	43.6
4	20.3	31.3	48.4	26.9	29.6	43.6
5 or more	16.9	26.9	56.1	18.6	33.2	48.2
Level of use						
Physician visits per person (59)						
1	34.2	9.1	56.7	30.8	18.5	50.6
2 to 4	21.7	26.0	52.3	27.3	29.5	43.3
5 or more	7.8	38.1	54.1	14.2	41.2	44.6
Expenditures for persons with physician office visits (24)						
$1 to 15	24.9	18.5	56.6	30.1	19.4	50.5
16 to 45	19.9	24.5	55.6	23.8	30.9	45.3
46 or more	10.3	35.8	54.0	8.2	47.9	43.9
Expenditures for persons with emergency room, outpatient, and clinic visits (14)						
$1 to 10	26.6	27.4	46.0	31.7	28.4	39.9
11 to 30	14.6	30.8	54.6	25.3	34.9	39.9
31 or more	9.3	38.0	52.7	32.7	34.9	41.0
Total	19.3%	27.4%	53.3%	23.5%	31.7%	44.8%

Table 6.6. Characteristics of Adults by Whether They Responded for Themselves or Had Others Respond for Them

Characteristic	*Respondent for Adult*		*Standard Error*
	Self	*Proxy*	
Demographic			
Age (6)			
17 to 25 years	33.9%	66.1%	2.2**
26 to 45	59.4	40.6	1.4**
46 to 55	63.9	36.1	1.9**
56 to 65	66.8	33.2	2.2**
66 years or more[a]	75.2	24.8	1.5
Family income (26)			
Nonpoor	55.4	44.8	1.1
Poor	67.7	32.3	1.2**
Race (62)			
White	58.3	41.7	1.1
Nonwhite	55.3	44.7	1.7
Residence (64)			
Rural nonfarm	56.5	43.5	1.4
Rural farm	57.5	42.5	3.0
SMSA central city	59.8	40.2	1.5
SMSA other urban[a]	56.0	44.0	2.0
Urban nonSMSA	59.7	40.3	3.8
Sex (68)			
Male	40.2	59.8	1.4
Female	74.0	26.0	1.0**
Sex and employment status (69)			
Male, employed full-time[a]	36.7	63.3	1.8
Male, not employed full-time	46.2	53.8	1.9**
Female, employed full-time	72.4	27.6	2.6**
Female, not employed full-time	73.4	26.6	1.2**
Perceived and evaluated health			
Perception of health (55)			
Excellent[a]	53.6	46.4	1.5
Good	57.9	42.1	1.3**
Fair	67.4	33.0	1.8**
Poor	66.8	33.2	3.2**
Worry about health (73)			
A great deal	60.8	39.2	2.1**
Some	61.6	38.4	2.0**
Hardly any	59.4	40.6	1.6**
None[a]	55.3	44.7	1.2

Table 6.6. (Continued)

Characteristic	*Respondent for Adult*		*Standard Error*
	Self	*Proxy*	
Severity over all diagnoses (67)			
Elective only[a]	52.4	47.6	1.7
Elective and mandatory	68.7	31.3	1.4**
Mandatory only	62.7	37.3	1.6**
Number of diagnoses (45)			
1[a]	55.7	44.3	1.5
2	59.6	40.4	2.4
3	71.0	29.0	2.2**
4 or more	72.4	27.6	2.6**
Total	57.9%	42.1%	1.0

Note: ** indicates that the difference between the compared categories is greater than two standard errors.

[a]Category against which others were compared to test the significance of the standard error of the difference.

Table 7.1. Noninterview Adjustment Factors When Adjustment is Based on Reason no Interview Obtained

	Degree of Uncooperativeness			
Number of Calls to Complete Main Questionnaire (67)[a]	*Uncooperative (broke appointments, refused to make them, would not open door, etc.)*	*Generally Cooperative, but Refused to Give Permission to Verify*	*Generally Cooperative, but Gave Only Oral Permission to Verify*	*Completely Cooperative*
1	2.3995	1.0153	1.0000	1.0000
2	2.3995	1.3416	1.0000	1.0000
3	2.3995	1.6443	1.0000	1.0000
4	2.3995	2.3995	2.3995	1.0053
5	2.3995	2.3995	2.3995	1.0151
6	2.3995	2.3995	2.3995	1.1076
7	2.3995	2.3995	2.3995	1.1148
8	2.3995	2.3995	2.3995	1.1616
9 or more	2.3995	2.3995	2.3995	2.3995

[a]In this and all subsequent tables in the chapter, the numbers in parentheses refer to the variable definition list found in Appendix C.

Table 7.2. Percentage Distribution of Sample Households by Noninterview Adjustment Factor

Information Used in Noninterview Adjustment	*Noninterview Adjustment Factor*						*Total*
	1.00	*Over 1.00 through 1.02*	*Over 1.02 through 1.15*	*Over 1.15 through 1.30*	*Over 1.30 through 2.00*	*Over 2.00 through 2.40*	
Geographic	15.8%	.0%	24.8%	34.1%	25.3%	.1%	100.0%
Reason no interview obtained	67.3	12.7	3.4	1.0	2.9	12.8	100.0

Table 7.3. Effect of Noninterview Adjustment on Distribution of Sample Persons

Characteristic	Percent of Weighted Sample Persons			
	Unadjusted Difference	*Estimates Adjusted for Nonresponse*		
		Adjusted Using External Data	*Adjusted Using Geographic Information*	*Adjusted Using Reason No Interview Obtained*
Demographic				
Age of oldest family member (7)				
Less than 65 years	86.2%	86.5%	86.2%	86.3%
65 years or more	13.8	13.4	13.8	13.7
Family income (26)				
Nonpoor	77.0	77.3	77.1	77.8
Poor	23.0	22.8	22.9	22.2
Race (62)				
White	87.9	87.9	88.0	88.3
Nonwhite	12.1	12.1	12.0	11.7
Residence (64)				
Rural nonfarm	24.5	23.7	24.4	24.4
Rural farm	6.8	6.2	6.8	6.6
SMSA central city	29.8	31.2	29.4	29.7
SMSA other urban	26.9	27.0	27.2	26.5
Urban nonSMSA	12.1	12.0	12.2	12.8
Perceived and evaluated health				
Perception of health (55)				
Excellent	37.8	—[a]	37.9	38.1
Good	43.0	—	42.7	42.9
Fair	12.4	—	12.4	12.1
Poor	3.9	—	3.9	3.9
Worry about health (73)				
A great deal	8.4	—	8.4	8.1
Some	18.2	—	18.3	18.3
Hardly any	20.6	—	20.6	20.2
None	50.1	—	50.1	50.6
Severity over all diagnoses (67)				
Elective only	23.9	—	23.9	24.6
Elective and mandatory	20.3	—	20.3	20.2
Mandatory only	20.2	—	20.2	19.7
Number of diagnoses (45)				
1	28.2	—	28.2	28.5
2	17.6	—	17.7	18.2
3	9.2	—	9.1	9.0
4 or more	9.0	—	9.0	8.4

Table 7.3. (Continued)

Characteristic	Percent of Weighted Sample Persons			
	Unadjusted Difference	*Estimates Adjusted for Nonresponse*		
		Adjusted Using External Data	*Adjusted Using Geographic Information*	*Adjusted Using Reason No Interview Obtained*
Level of use				
Physician visits per person (59)				
0	32.4	—	32.4	32.1
1	18.6	—	18.5	19.0
2 to 4	24.1	—	24.1	24.1
5 or more	24.9	—	25.0	24.8
5 to 13	18.7	—	18.7	18.9
14 or more	6.2	—	6.3	5.9
Hospital admissions per person (31)				
0	88.7	—	88.7	88.9
1	9.0	—	9.1	8.7
2 or more	2.3	—	2.3	2.4
Total	100.0%	100.0%	100.0%	100.0%

[a]Data necessary to provide these estimates were unavailable.

Table 7.4. Effect of Nonresponse Adjustment on Estimates of Mean Number of Physician Visits for Persons Seeing a Physician

Characteristic	*Physician Visits per Person (59)*			
	Unadjusted Social Survey Estimate	*Estimates Adjusted for Nonresponse*		
		Adjusted Using External Data	*Adjusted Using Geographic Information*	*Adjusted Using Reason No Interview Obtained*
Demographic				
Age of oldest family member (7)				
Less than 65 years	5.4	5.5	5.4	5.3
65 years or more	7.8	7.9	7.8	7.7
Family income (26)				
Nonpoor	5.5	5.6	5.6	5.5
Poor	6.5	6.6	6.6	6.5
Race (62)				
White	5.7	5.7	5.7	5.6
Nonwhite	6.0	6.1	5.9	5.9
Residence (64)				
Rural nonfarm	5.4	5.4	5.4	5.3
Rural farm	5.6	5.6	5.6	5.7
SMSA central city	6.0	6.1	6.1	6.0
SMSA other urban	5.6	5.6	5.6	5.4
Urban nonSMSA	6.1	6.2	6.2	6.1
Perceived and evaluated health				
Perception of health (55)				
Excellent	3.6	—[a]	3.6	3.5
Good	5.3	—	5.3	5.2
Fair	9.1	—	9.2	8.9
Poor	14.2	—	14.2	14.5
Worry about health (73)				
A great deal	11.7	—	11.8	11.5
Some	7.4	—	7.5	7.3
Hardly any	4.9	—	5.0	4.9
None	3.6	—	3.6	3.6
Severity over all diagnoses (67)				
Elective only	3.2	—	3.2	3.2
Elective and mandatory	8.3	—	8.4	8.1
Mandatory only	7.2	—	7.2	7.3
Number of diagnoses (45)				
1	3.8	—	3.8	3.8
2	5.6	—	5.7	5.8
3	8.0	—	8.0	7.8
4 or more	11.6	—	11.7	11.6

Table 7.4. (Continued)

Characteristic	Physician Visits per Person (59)			
	Unadjusted Social Survey Estimate	*Estimates Adjusted for Nonresponse*		
		Adjusted Using External Data	*Adjusted Using Geographic Information*	*Adjusted Using Reason No Interview Obtained*
Level of use				
Mean number of visits (59)				
1	1.0	—	1.0	1.0
2 to 4	2.8	—	2.8	2.8
5 or more	12.1	—	12.2	12.1
5 to 13	8.0	—	8.0	8.1
14 or more	24.8	—	24.7	24.9
Total	5.7	5.8	5.8	5.7

[a]Data necessary to provide these estimates were unavailable.

Table 7.5. Effect of Nonresponse Adjustment on Estimates of Mean Hospital Expenditures per Admission

Characteristic	Expenditures per Admission (23)			
	Unadjusted Social Survey Estimate	Estimates Adjusted for Nonresponse		
		Adjusted Using External Data	Adjusted Using Geographic Information	Adjusted Using Reason No Interview Obtained
Demographic				
Age of oldest family member (7)				
Less than 65 years	$ 640	$662	$ 642	$ 624
65 years or more	863	886	877	857
Family income (26)				
Nonpoor	674	697	679	643
Poor	715	737	717	749
Race (62)				
White	684	709	691	659
Nonwhite	688	711	669	770
Residence (64)				
Rural nonfarm	557	573	565	534
Rural farm	575	586	562	574
SMSA central city	727	757	745	735
SMSA other urban	910	941	904	904
Urban nonSMSA	444	458	446	431
Perceived and evaluated health				
Perception of health (55)				
Excellent	488	—[a]	495	475
Good	609	—	619	614
Fair	698	—	712	682
Poor	868	—	877	862
Worry about health (73)				
A great deal	1,036	—	1,049	978
Some	560	—	567	544
Hardly any	442	—	440	506
None	441	—	437	449
Severity over all diagnoses (67)				
Elective only	499	—	501	482
Elective and mandatory	550	—	554	539
Mandatory only	830	—	833	807
Number of diagnoses (45)				
1	626	—	643	617
2	573	—	578	620
3	563	—	559	544
4 or more	883	—	886	839

Table 7.5. (Continued)

Characteristic	Expenditures per Admission (23)			
	Unadjusted Social Survey Estimate	Estimates Adjusted for Nonresponse		
		Adjusted Using External Data	Adjusted Using Geographic Information	Adjusted Using Reason No Interview Obtained
Level of use				
Expenditures per admission (23)				
$0 to 300	184	—	183	182
301 to 600	426	—	426	423
601 or more	1,556	—	1,574	1,480
601 to 1200	843	—	841	845
1201 or more	2,801	—	2,812	2,704
Total	$ 685	$707	$ 689	$ 670

[a]Data necessary to provide these estimates were unavailable.

Table 7.6. Effect of Nonresponse Adjustment on Differences between Groups for Mean Number of Physician Visits for Persons Seeing a Physician

Characteristic	Physician Visits per Person (59)			
	Unadjusted Difference	Difference Adjusted for Nonresponse		
		Adjusted Using External Data	Adjusted Using Geographic Information	Adjusted Using Reason No Interview Obtained
Demographic				
Age of oldest family member (7)				
Less than 65 versus 65 or more	– 2.4(6.6)[a]	–2.5(6.6)	– 2.4(6.2)	– 2.4(5.9)
Family income (26)				
Nonpoor versus poor	– 1.0(3.7)	–1.0(3.8)	– 1.1(3.7)	– 1.0(3.3)
Race (62)				
White versus nonwhite	– .3(.6)	– .3(.7)	– .2(.4)	– .3(.5)
Residence (64)				
SMSA other urban versus:				
Rural nonfarm	.2(.4)	.1(.4)	.2(.6)	.1(.3)
Rural farm	.0(.0)	.0(.0)	.0(.1)	– .3(.4)
SMSA central city	– .5(1.1)	– .5(1.2)	– .5(1.1)	– .6(1.3)
Urban nonSMSA	– .6(1.1)	– .6(1.0)	– .6(1.2)	– .7(1.0)
Perceived and evaluated health				
Perception of health (55)				
Excellent versus:				
Good	– 1.7(6.7)	—[b]	– 1.8(6.8)	– 1.7(2.3)
Fair	– 5.5(10.8)	—	– 5.6(11.0)	– 5.4(11.2)
Poor	–10.6(9.3)	—	–10.6(8.7)	–11.0(6.8)

Worry about health (73)				
None versus:				
Hardly any	– 1.3(4.4)	—	– 1.4(4.5)	– 1.3(3.7)
Some	– 3.8(10.2)	—	– 3.8(10.1)	– 3.6(9.8)
A great deal	– 8.1(11.3)	—	– 8.2(10.0)	– 7.9(9.9)
Severity over all diagnoses (67)				
Elective only versus:				
Elective and mandatory	– 5.1(15.9)	—	– 5.2(15.7)	– 5.0(15.1)
Mandatory only	– 4.0(11.6)	—	– 4.1(11.6)	– 4.1(10.8)
Number of diagnoses (45)				
1 versus:				
2	– 1.8(5.2)	—	– 1.9(5.2)	– 2.0(4.5)
3	– 4.1(9.6)	—	– 4.2(10.0)	– 4.0(10.6)
4 or more	– 7.8(20.6)	—	– 7.9(20.8)	– 7.9(19.2)
Level of use				
Physician visits per person (59)				
1 versus:				
2 to 4	– 1.8(83.1)	—	– 1.8(93.8)	– 1.8(71.5)
5 or more	–11.1(39.5)	—	–11.2(39.4)	–11.1(33.1)
5 to 13	– 7.0(72.8)	—	– 7.0(72.9)	– 7.1(60.8)
14 or more	–23.8(29.3)	—	–23.7(27.7)	–23.9(25.8)

[a]Numbers in parentheses in the body of the table are the ratios of the differences to the estimated standard errors of the differences.
[b]Data necessary to provide these estimates were unavailable.

Table 7.7. Effect of Nonresponse Adjustment on Differences between Groups for Mean Hospital Expenditures per Admission

Characteristic	Expenditures per Admission (23)			
	Unadjusted Difference	Difference Adjusted for Nonresponse		
		Adjusted Using External Data	Adjusted Using Geographic Information	Adjusted Using Reason No Interview Obtained
Demographic				
Age of oldest family member (7)				
Less than 65 versus				
65 or more	$− 222(3.0)[a]	$− 225(3.0)	$− 235(3.0)	$− 233(3.3)
Family income (26)				
Nonpoor versus poor	− 41(.6)	− 40(.6)	− 37(.5)	− 106(1.3)
Race (62)				
White versus nonwhite	− 3(.0)	− 2(.0)	22(.3)	− 111(.7)
Residence (64)				
SMSA other urban versus:				
Rural nonfarm	353(2.2)	368(2.2)	339(2.1)	371(2.4)
Rural farm	335(2.1)	355(2.2)	342(2.1)	330(2.1)
SMSA central city	183(1.2)	184(1.2)	160(1.0)	169(1.1)
Urban nonSMSA	467(2.9)	483(3.0)	458(2.9)	473(3.1)
Perceived and evaluated health				
Perception of health (55)				
Excellent versus:				
Good	− 122(1.1)	—[b]	− 124(1.0)	− 139(1.5)
Fair	− 210(2.1)	—	− 217(1.9)	− 207(2.3)
Poor	− 380(3.6)	—	− 382(3.2)	− 387(3.8)

Worry about health (73)				
None versus:				
Hardly any	– 1(.0)	—	– 3(.0)	– 58(.7)
Some	– 119(1.6)	—	– 129(1.6)	– 95(.7)
A great deal	– 595(6.6)	—	– 612(6.8)	– 530(5.8)
Severity over all diagnoses (67)				
Elective only versus:				
Elective and mandatory	– 51(.6)	—	– 54(.6)	– 57(.7)
Mandatory only	– 331(3.5)	—	– 333(3.6)	– 325(3.7)
Number of diagnoses (45)				
1 versus:				
2	53(.5)	—	64(.6)	– 3(.0)
3	63(.7)	—	83(.8)	72(.9)
4 or more	– 257(2.2)	—	– 243(2.0)	– 222(2.2)
Level of use				
Expenditures per admission (23)				
$0 to 300 versus:				
301 to 600	– 242(36.2)	—	– 243(35.9)	– 241(33.5)
601 or more	– 1,371(12.2)	—	– 1,391(12.6)	– 1,298(11.6)
601 to 1200	– 659(35.3)	—	– 658(34.2)	– 662(32.6)
1201 or more	– 2,616(11.1)	—	– 2,539(11.8)	– 2,522(9.9)

[a]Numbers in parentheses in the body of the table are the ratios of the differences to the estimated standard errors of the differences.
[b]Data necessary to provide these estimates were unavailable.

Table 8.1. Imputed Social Survey Expenditures for Hospital Admissions

Characteristic	Percent of Admissions with Imputed Expenditures	Unweighted Admissions with Imputed Expenditures		Percent of Expenditures Imputed
		Percent	N	
Demographic				
Age of oldest family member (7)[a]				
Less than 65 years	28%	39%	493	32%
65 years or more	46	49	241	45
Family income (26)				
Nonpoor	25	31	297	27
Poor	51	55	437	56
Race (62)				
White	30	35	441	32
Nonwhite	52	57	293	66
Residence (64)				
Rural nonfarm	29	29	129	33
Rural farm	23	23	39	32
SMSA central city	37	53	406	32
SMSA other urban	28	39	90	43
Urban nonSMSA	42	44	70	39
Perceived and evaluated health				
Perception of health (55)				
Excellent	25	31	84	24
Good	24	34	220	24
Fair	42	47	215	47
Poor	44	51	137	56
Worry about health (73)				
A great deal	36	45	277	35
Some	28	40	200	29
Hardly any	30	40	94	46
None	44	36	127	33
Severity over all diagnoses (67)				
Elective only	42	45	43	55
Elective and mandatory	27	36	240	28
Mandatory only	36	45	435	38
Number of diagnoses (45)				
1	34	44	203	34
2	30	40	169	44
3	28	41	141	40
4 or more	34	41	206	30

Table 8.1. (Continued)

Characteristic	Percent of Admissions with Imputed Expenditures	Unweighted Admissions with Imputed Expenditures		Percent of Expenditures Imputed
		Percent	N	
Level of use				
Expenditures per admission (23)				
$0 to 300	32	38	249	32
301 to 600	31	42	231	32
601 or more	33	46	254	34
Total	32%	42%	734	35%

Note: All hospital expenditures (weighted) = $5,492,054. Total unweighted social survey admissions = 1,769. Total weighted social survey admissions = 8,021.

[a]In this and all subsequent tables in the chapter, the numbers in parentheses refer to the variable definition list found in Appendix C.

Table 8.2. Effect of Item Imputation on Estimates of Hospital Expenditures per Admission

Characteristic	Expenditures per Admission (23)			
	Using External Item Estimation	Using Internal Item Imputation		
		Cell Mean Method	Hot Deck Method	
			PSU Numbers Ordered Low to High	PSU Numbers Ordered High to Low
Demographic				
Age of oldest family member (7)				
Less than 65 years	$ 640	$ 654	$ 652	$ 638
65 years or more	863	973	932	931
Family income (26)				
Nonpoor	674	708	697	684
Poor	715	745	736	730
Race (62)				
White	684	721	715	700
Nonwhite	688	687	629	658
Residence (64)				
Rural nonfarm	557	582	567	566
Rural farm	575	634	611	642
SMSA central city	727	761	754	747
SMSA other urban	910	933	930	873
Urban nonSMSA	444	495	480	516
Perceived and evaluated health				
Perception of health (55)				
Excellent	488	496	502	483
Good	609	623	614	600
Fair	698	785	759	762
Poor	868	929	861	910
Worry about health (73)				
A great deal	1,036	1,097	1,076	1,072
Some	560	602	593	579
Hardly any	442	433	419	413
None	441	442	445	437
Severity over all diagnoses (67)				
Elective only	499	484	521	462
Elective and mandatory	550	575	558	555
Mandatory only	830	875	866	853
Number of diagnoses (45)				
1	626	692	703	641
2	573	551	561	527
3	563	606	573	619
4 or more	883	924	896	908

Table 8.2. (Continued)

Characteristic	Expenditures per Admission (23)			
	Using External Item Estimation	Using Internal Item Imputation		
		Cell Mean Method	Hot Deck Method	
			PSU Numbers Ordered Low to High	PSU Numbers Ordered High to Low
Level of use				
Expenditures per admission (23)				
$0 to 300	184	220	225	209
301 to 600	426	447	443	431
601 or more	1,556	1,599	1,562	1,561
601 to 1200	843	874	856	867
1201 or more	2,801	2,865	2,796	2,773
Total	$ 685	$ 717	$ 707	$ 696

Table 8.3. Effect of Item Imputation on Estimates of Differences of Hospital Expenditures per Admission

Characteristic	*Expenditures per Admission (23)*			
	Using External Item Estimation	*Using Internal Item Imputation*		
		Cell Mean Method	*Hot Deck Method*	
			PSU Numbers Ordered Low to High	*PSU Numbers Ordered High to Low*
Demographic				
Age of oldest family member (7)				
Less than 65 versus 65 or more	$– 222(3.0)	$– 319(3.8)	$– 281(3.4)	$– 294(3.7)
Family income (26)				
Nonpoor versus poor	– 41(.6)	– 37(.5)	– 39(.6)	– 46(.7)
Race (62)				
White versus nonwhite	– 3(.0)	34(.5)	86(1.5)	42(.7)
Residence (64)				
SMSA other urban versus:				
Rural nonfarm	353(2.2)	350(2.2)	362(2.2)	306(1.9)
Rural farm	335(2.1)	298(1.7)	318(1.9)	230(1.3)
SMSA central city	183(1.2)	171(1.1)	176(1.2)	126(.8)
Urban nonSMSA	467(2.9)	437(2.9)	450(2.9)	357(2.2)
Perceived and evaluated health				
Perception of health (55)				
Excellent versus:				
Good	– 122(1.1)	– 127(1.2)	– 112(1.1)	– 118(1.1)
Fair	– 210(2.1)	– 289(2.6)	– 257(2.5)	– 279(2.4)
Poor	– 380(3.6)	– 433(3.6)	– 360(3.2)	– 428(3.5)

Worry about health (73)				
None versus:				
Hardly any	– 1(.0)	9(.2)	26(.6)	24(.4)
Some	– 119(1.6)	– 159(2.2)	– 148(2.2)	– 142(1.8)
A great deal	– 595(6.6)	– 655(7.2)	– 631(6.9)	– 416(4.5)
Severity over all diagnoses (67)				
Elective only versus:				
Elective and mandatory	– 51(.6)	– 91(1.3)	– 38(.5)	– 93(1.2)
Mandatory only	– 331(3.5)	– 391(4.6)	– 345(3.7)	– 391(4.3)
Number of diagnoses (45)				
1 versus:				
2	53(.5)	141(1.5)	141(1.5)	115(1.2)
3	63(.7)	86(.9)	130(1.3)	22(.2)
4 or more	– 257(2.2)	– 232(1.9)	– 193(1.6)	– 267(2.2)
Level of use				
Expenditures per admission (23)				
$0 to 300 versus:				
301 to 600	– 242(36.2)	– 227(19.6)	– 219(13.4)	– 222(18.3)
601 or more	– 1,371(12.2)	– 1,379(12.8)	– 1,337(12.7)	– 1,352(11.9)
601 to 1,200	– 659(35.3)	– 654(23.1)	– 631(20.5)	– 658(25.2)
1,201 or more	– 2,616(11.1)	– 2,646(12.2)	– 2,571(10.5)	– 2,565(12.1)

Table 9.1. Initial and Intermediate Weights and their Corresponding Number of Interviewed Households

	Sample			
Category	*U*	*A*	*R*	*S*
Weights				
Weight based on selection probability	1.00000	13.86232	7.21427	8.11153
Weight adjusted by differential nonresponse	1.00000	13.96462	6.87833	7.99441
Intermediate weight				
U-type urban low income (26)[a] and/or old (7)	1.00000	—	—	1.00000
U-type urban other	1.12505	—	—	—
A-type urban low income (26) and/or old (7)	—	5.71978	—	5.71978
A-type urban other	—	15.71088	—	—
Rural low income (26) and/or old (7)	—	3.28894	3.28894	3.28894
Rural other	—	5.18478	5.18478	—
Unweighted number of interviewed households				
U-type urban low income (26) and/or old (7)	748	—	—	108
U-type urban other	628	—	—	—
A-type urban low income (26) and/or old (7)	—	275	—	458
A-type urban other	—	544	—	—
Rural low income (26) and/or old (7)	—	132	242	218
Rural other	—	168	359	—

[a]In this and all subsequent tables in the chapter, the numbers in parentheses refer to the variable definition list found in Appendix C.

Table 9.2. Comparison of the Size and Composition of the Weighted Sample and the Sample Adjusted as Though Self-Weighting

Characteristic	*Percent of Sample Persons*		*Number of Sample Persons*		*Ratio of Number of Sample Persons*
	Weighted Sample[a]	*Self-Weighting Sample*[b]	*Weighted Sample*[a]	*Self-Weighting Sample*[b]	
Demographic					
Age of oldest family member (7)					
Less than 65 years	81.3%	86.2%	9,451	10,630	.89
65 years or more	18.7	13.8	2,168	1,697	1.28
Family income (26)					
Nonpoor	57.2	77.0	6,646	9,487	.70
Poor	42.8	23.0	4,973	2,840	1.75
Race (62)					
White	67.6	87.9	7,855	10,841	.72
Nonwhite	32.4	12.1	3,764	1,486	2.53
Residence (64)					
Rural nonfarm	24.1	24.5	2,799	3,014	.93
Rural farm	9.2	6.9	1,069	845	1.26
SMSA central city	46.0	29.8	5,350	3,669	1.46
SMSA other urban	13.4	26.9	1,560	3,313	.47
Urban nonSMSA	7.2	12.1	841	1,493	.56
Perceived and evaluated health					
Perception of health (55)					
Excellent	33.7	37.8	3,910	4,654	.84
Good	43.1	43.0	5,005	5,299	.94
Fair	14.6	12.4	1,698	1,523	1.12
Poor	5.1	3.9	590	475	1.24
Worry about health (73)					
A great deal	9.6	8.4	1,117	1,033	1.08
Some	17.7	18.2	2,057	2,244	.92
Hardly any	19.5	20.6	2,265	2,542	.89
None	50.1	50.1	5,824	6,176	.94
Severity over all diagnoses (67)					
Elective only	20.2	23.9	2,349	2,943	.80
Elective and mandatory	17.4	20.3	2,022	2,503	.81
Mandatory only	20.4	20.2	2,369	2,486	.95
Number of diagnoses (45)					
1	26.0	28.2	3,024	3,472	.87
2	15.3	17.6	1,772	2,174	.82
3	8.3	9.2	970	1,129	.86
4 or more	8.0	9.0	934	1,104	.85

Table 9.2. (Continued)

Characteristic	Percent of Sample Persons		Number of Sample Persons		Ratio of Number of Sample Persons
	Weighted Sample[a]	Self-Weighting Sample[b]	Weighted Sample[a]	Self-Weighting Sample[b]	
Level of use					
Physician visits per person (59)					
0	38.6	32.4	4,487	3,988	1.12
1	17.0	18.6	1,978	2,309	.86
2 to 4	20.9	24.1	2,429	2,967	.82
5 or more	23.5	24.9	2,725	3,063	.89
5 to 13	17.3	18.7	2,006	2,302	.87
14 or more	6.2	6.2	719	762	.94
Hospital admissions per person (31)					
0	88.0	88.7	10,228	10,929	.94
1	9.3	9.0	1,082	1,110	.98
2 or more	2.7	2.3	309	289	1.07
Expenditures per person with admission (19)					
$0 to 300	3.7	3.2	432	392	1.10
301 to 600	3.6	4.0	417	491	.85
601 or more	4.7	4.1	541	511	1.06
Total	100.0%	100.0%	11,619	12,327	.94

[a]This column reflects the distribution of persons interviewed in the weighted sample. Since various types of persons were o' rsampled in this survey, the distribution of persons interviewed does not necessarily reflect the relative number of Americans in each category.

[b]This column reflects the expected relative number of persons in each category in the civilian noninstitutionalized population.

Table 9.3. Further Comparison of the Weighted Sample and Expected Results from a Self-Weighting Sample

Characteristic	*Physician Visits per Person (59)*				*Expenditures per Admission (23)*			
	Weighted Sample		*Sample Adjusted as Though Self-Weighting*		*Weighted Sample*		*Sample Adjusted as Though Self-Weighting*	
	Number of Persons Seeing Physician	*Standard Error of Mean Number of Visits*	*Number of Persons Seeing Physician*	*Standard Error of Mean Number of Visits*	*Number of Hospital Admissions*	*Standard Error of Mean Expenditure*	*Number of Hospital Admissions*	*Standard Error of Mean Expenditure*
Demographic								
Age of oldest family member (7)								
Less than 65 years	5,698	.1565	7,158	.1047	1,279	47.04	1,393	37.08
65 years or more	1,434	.3401	1,182	.3397	490	58.17	347	68.57
Family income (26)								
Nonpoor	4,445	.1788	6,706	.1162	972	43.55	1,291	27.88
Poor	2,687	.2057	1,633	.2465	797	55.18	449	63.97
Race (62)								
White	5,179	.1705	7,478	.1087	1,259	39.77	1,572	26.76
Nonwhite	1,953	.4501	861	.5404	510	81.90	169	94.67
Residence (64)								
Rural nonfarm	1,817	.2452	2,041	.1821	443	65.77	446	47.22
Rural farm	644	.5593	503	.6720	166	62.58	120	75.46
SMSA central city	3,060	.3249	2,383	.2526	772	40.94	453	38.06
SMSA other urban	1,034	.2708	2,365	.1550	230	150.30	481	85.10
Urban nonSMSA	577	.4648	1,048	.3151	158	53.29	240	39.71

Table 9.3. (Continued)

Characteristic	*Physician Visits per Person (59)*				*Expenditures per Admission (23)*			
	Weighted Sample		*Sample Adjusted as Though Self-Weighting*		*Weighted Sample*		*Sample Adjusted as Though Self-Weighting*	
	Number of Persons Seeing Physician	*Standard Error of Mean Number of Visits*	*Number of Persons Seeing Physician*	*Standard Error of Mean Number of Visits*	*Number of Hospital Admissions*	*Standard Error of Mean Expenditure*	*Number of Hospital Admissions*	*Standard Error of Mean Expenditure*
Perceived and evaluated health								
Perception of health (55)								
Excellent	2,148	.1845	2,863	.1115	269	83.78	335	46.53
Good	2,914	.1819	3,573	.1181	648	67.90	723	44.20
Fair	1,270	.4709	1,194	.3173	453	58.64	349	62.20
Poor	497	1.1222	417	.8590	267	64.36	221	83.48
Worry about health (73)								
A great deal	934	.6827	928	.5127	621	82.31	598	56.33
Some	1,650	.3031	1,919	.2007	499	66.53	506	41.69
Hardly any	1,444	.2072	1,758	.1421	236	64.57	246	60.16
None	2,845	.2202	3,491	.1327	351	35.11	347	58.20
Severity over all diagnoses (67)								
Elective only	2,318	.2005	2,915	.1133	95	72.56	89	159.19
Elective and mandatory	1,995	.2444	2,478	.1683	671	45.63	772	28.99
Mandatory only	2,224	.2811	2,349	.2282	970	61.42	861	45.47
Number of diagnoses (45)								
1	2,940	.1760	3,400	.1158	466	83.91	437	53.26
2	1,733	.3046	2,142	.1994	425	51.22	400	58.84
3	957	.3895	1,118	.2636	344	46.24	316	74.70
4 or more	917	.3326	1,092	.2710	503	81.71	570	52.61

Level of Use								
Physician visits per person (59)								
1	1,978	0	2,296	0	—	—	—	—
2 to 4	2,429	.0211	2,976	.0136	—	—	—	—
5 or more	2,725	.2823	3,067	.2087	—	—	—	—
5 to 13	2,006	.0956	2,304	.0622	—	—	—	—
14 or more	719	.8104	763	.5630	—	—	—	—
Expenditures per admission (23)								
$0 to 300	—	—	—	—	663	4.58	609	3.47
301 to 600	—	—	—	—	552	4.86	602	3.34
601 or more	—	—	—	—	554	112.22	528	72.71
601 to 1200	—	—	—	—	334	18.09	336	12.02
1201 or more	—	—	—	—	220	236.32	192	145.47
Total	7,132	.1602	8,339	.1134	1,769	38.77	1,741	28.59

Table 9.4. Original Post-Stratification Adjustment Categories

Characteristic				*Post-Stratification Adjustment Factor*	*Percent of Sample Households*	
Race (62)	*Residence (64)*	*Family Size (27)*	*Family Income (26)*		*Unweighted, Unadjusted*	*Weighted, Adjusted*
White	SMSA	1	Under $3,000	1.622	6.0%	5.9%
White	SMSA	1	$3,000 and over	1.191	5.6	8.0
White	SMSA	2+	Under $3,000	1.136	3.3	2.5
White	SMSA	2+	$3,000-14,999	1.160	18.6	29.8
White	SMSA	2+	$15,000 and over	1.181	5.6	12.3
White	NonSMSA	1	All categories	.967	5.9	5.7
White	NonSMSA	2+	Under $3,000	1.080	3.3	2.7
White	NonSMSA	2+	$3,000-14,999	.750	21.9	18.4
White	NonSMSA	2+	$15,000 and over	1.147	2.4	3.8
Nonwhite	SMSA	1	Under $3,000	1.200	3.3	1.2
Nonwhite	SMSA	1	$3,000 and over	.714	2.2	1.0
Nonwhite	SMSA	2+	Under $3,000	.714	4.3	1.0
Nonwhite	SMSA	2+	$3,000-14,999	.600	13.6	4.1
Nonwhite	SMSA	2+	$15,000 and over	1.167	.8	.7
Nonwhite	NonSMSA	1	All categories	1.000	.6	.6
Nonwhite	NonSMSA	2+	All categories	.917	2.6	2.2
Total				1.000	100.0%	100.0%

Table 9.5. Alternative Post-Stratification Adjustment Categories

Characteristic			*Post-Stratification Adjustment Factor*	*Percent of Sample Persons*	
Race (62)	*Sex (68)*	*Age (6)*		*Unweighted, Unadjusted*	*Weighted, Adjusted*
White	Both	Less than 6	1.0344	6.5%	8.7%
White	Both	6 to 11	.9699	7.9	10.2
White	Both	12 to 17	.9563	8.3	10.2
White	Male	18 to 29	1.1487	5.0	7.8
White	Female	18 to 29	1.1417	5.5	8.3
White	Male	30 to 44	1.0336	5.1	7.3
White	Female	30 to 44	1.0036	5.4	7.5
White	Both	45 to 54	1.1228	6.7	10.3
White	Both	55 to 64	1.0330	6.6	8.3
White	Both	65 to 74	1.0245	6.5	5.5
White	Both	75 or more	1.1384	4.2	3.5
Nonwhite	Both	Less than 9	.7906	7.3	2.8
Nonwhite	Both	9 to 17	.6803	8.4	2.6
Nonwhite	Male	18 to 44	.9781	3.8	2.0
Nonwhite	Female	18 to 44	.7465	6.0	2.3
Nonwhite	Both	45 to 64	.7459	4.7	2.0
Nonwhite	Both	65 or more	.7410	2.3	.8
Total			1.0000	100.0%	100.0%

Table 9.6. Effect of Post-Stratification Adjustment on Distribution of Sample Persons

	Percent of Weighted Sample Persons		
	Without Post-Stratification Adjustment	*With Post-Stratification Adjustment*	
Characteristic		*Original Categories*	*Alternative Categories*
Demographic			
Age of oldest family member (7)			
Less than 65 years	86.8%	86.2%	86.5%
65 years or more	13.3	13.8	13.5
Family income (26)			
Nonpoor	75.7	77.0	77.1
Poor	24.3	23.0	22.9
Race (62)			
White	83.8	87.9	87.5
Nonwhite	16.2	12.1	12.5
Residence (64)			
Rural nonfarm	25.8	24.5	26.5
Rural farm	7.8	6.8	7.9
SMSA central city	29.1	29.8	27.7
SMSA other urban	23.3	26.9	23.8
Urban nonSMSA	13.9	12.1	14.2
Perceived and evaluated health			
Perception of health (55)			
Excellent	37.0	37.8	37.4
Good	43.3	43.0	43.1
Fair	12.7	12.4	12.7
Poor	3.8	3.9	3.8
Worry about health (73)			
A great deal	8.3	8.4	8.3
Some	18.2	18.2	18.4
Hardly any	20.7	20.6	20.8
None	50.1	50.1	49.8
Severity over all diagnoses (67)			
Elective only	23.4	23.9	23.4
Elective and mandatory	19.8	20.3	20.3
Mandatory only	20.1	20.2	20.3
Number of diagnoses (45)			
1	27.9	28.2	28.0
2	17.5	17.6	17.6
3	8.9	9.2	9.2
4 or more	8.6	9.0	8.9

Table 9.6. (Continued)

Characteristic	*Percent of Weighted Sample Persons*		
	Without Post-Stratification Adjustment	*With Post-Stratification Adjustment*	
		Original Categories	*Alternative Categories*
Level of use			
Physician visits per person (59)			
0	33.1	32.4	32.6
1	18.5	18.6	18.4
2 to 4	23.9	24.1	24.1
5 or more	24.5	24.9	24.9
5 to 13	18.4	18.7	18.7
14 or more	6.1	6.2	6.2
Hospital admissions per person (31)			
0	88.6	88.7	88.3
1	9.1	9.0	9.3
2 or more	2.3	2.3	2.4
Total	100.0%	100.0%	100.0%

Table 9.7. Effect of Post-Stratification Adjustment on Estimates of Physician Visits for Persons Seeing a Physician and on Estimates of Hospital Expenditures per Admission

Characteristic	*Physician Visits per Person (59)*			*Expenditures per Admission (23)*		
	Without Post-Stratification Adjustment	*With Post-Stratification Adjustment*		*Without Post-Stratification Adjustment*	*With Post-Stratification Adjustment*	
		Original Categories	*Alternative Categories*		*Original Categories*	*Alternative Categories*
Demographic						
Age of oldest family member (7)						
Less than 65 years	5.4	5.4	5.4	$ 616	$ 640	$ 616
65 years or more	7.7	7.8	7.7	830	863	831
Family income (26)						
Nonpoor	5.5	5.5	5.6	649	674	647
Poor	6.3	6.5	6.3	685	715	691
Race (62)						
White	5.7	5.7	5.7	652	684	650
Nonwhite	5.9	6.0	5.9	702	688	730
Residence (64)						
Rural nonfarm	5.4	5.4	5.4	544	557	540
Rural farm	5.5	5.6	5.5	555	575	566
SMSA central city	6.0	6.0	6.0	706	727	711
SMSA other urban	5.5	5.6	5.5	920	910	915
Urban nonSMSA	6.1	6.1	6.2	447	444	447
Perceived and evaluated health						
Perception of health (55)						
Excellent	3.5	3.6	3.6	467	488	472
Good	5.2	5.3	5.3	600	609	596
Fair	8.9	9.1	9.0	670	698	682
Poor	14.1	14.2	14.2	816	868	809

Worry about health (73)						
A great deal	11.6	11.7	11.6	973	1,036	980
Some	7.3	7.4	7.3	539	560	537
Hardly any	4.9	4.9	5.0	487	442	481
None	3.6	3.6	3.7	427	441	427
Severity over all diagnoses (67)						
Elective only	3.1	3.2	3.2	472	499	473
Elective and mandatory	8.3	8.3	8.3	536	550	538
Mandatory only	7.1	7.2	7.1	789	830	792
Number of diagnoses (45)						
1	3.8	3.8	3.8	605	626	600
2	5.6	5.6	5.7	573	573	574
3	8.1	8.0	8.0	543	563	540
4 or more	11.6	11.6	11.5	836	883	839
Level of use						
Physician visits per person (59)						
1	1.0	1.0	1.0	—	—	—
2 to 4	2.7	2.8	2.7	—	—	—
5 or more	12.1	12.1	12.1	—	—	—
5 to 13	8.0	8.0	8.0	—	—	—
14 or more	24.8	24.8	24.7	—	—	—
Expenditures per admission (23)						
$0 to 300	—	—	—	184	184	185
301 to 600	—	—	—	427	426	427
601 or more	—	—	—	1,525	1,556	1,525
601 to 1200	—	—	—	841	843	842
1201 or more	—	—	—	2,769	2,801	2,771
Total	5.7	5.7	5.7	$ 658	$ 685	$ 658

Table 9.8. Effect of Post-Stratification Adjustment on Differences Between Groups for Mean Number of Physician Visits for Persons Seeing a Physician and on Differences Between Groups for Hospital Expenditures per Admission

Characteristic	*Physician Visits per Person (59)*			*Expenditures per Admission (23)*		
	Without Post-Stratification Adjustment	*With Post-Stratification Adjustment*		*Without Post-Stratification Adjustment*	*With Post-Stratification Adjustment*	
		Original Categories	*Alternative Categories*		*Original Categories*	*Alternative Categories*
Demographic						
Age of oldest family member (7)						
Less than 65 versus 65 or more	− 2.4(6.4)[a]	− 2.4(6.6)	− 2.3(6.2)	$− 214(3.2)	$− 222(3.0)	$− 215(3.1)
Family income (26)						
Nonpoor versus poor	− .8(3.1)	− 1.0(3.7)	− .7(2.9)	− 36(.5)	− 41(.6)	− 43(.6)
Race (62)						
White versus nonwhite	− .2(.5)	− .3(.6)	− .2(.4)	− 50(.5)	− 3(.0)	− 80(.7)
Residence (64)						
SMSA other urban versus:						
Rural nonfarm	.2(.5)	.2(.4)	.2(.4)	377(2.3)	353(2.2)	375(2.3)
Rural farm	.0(.0)	.0(.0)	.1(.1)	365(2.2)	335(2.1)	349(2.1)
SMSA central city	− .4(1.1)	− .5(1.1)	− .5(1.2)	214(1.3)	183(1.2)	205(1.3)
Urban nonSMSA	− .6(1.2)	− .6(1.1)	− .7(1.2)	474(2.9)	467(2.9)	467(2.9)
Perceived and evaluated health						
Perception of health (55)						
Excellent versus:						
Good	− 1.7(6.9)	− 1.7(6.7)	− 1.7(7.0)	− 132(1.4)	− 122(1.1)	− 123(1.2)
Fair	− 5.4(11.7)	− 5.5(10.8)	− 5.4(11.8)	− 203(2.3)	− 210(2.1)	− 209(2.2)
Poor	−10.6(9.7)	−10.6(9.3)	−10.6(9.2)	− 348(3.7)	− 380(3.6)	− 337(3.5)

Worry about health (73)						
None versus:						
Hardly any	– 1.3(4.0)	– 1.3(4.4)	– 1.3(3.9)	– 60(.6)	– 1(.0)	– 54(.6)
Some	– 3.7(9.9)	– 3.8(10.2)	– 3.6(9.5)	– 112(1.7)	– 119(1.6)	– 110(1.7)
A great deal	– 8.0(12.1)	– 8.1(11.3)	– 8.0(11.8)	– 547(7.1)	– 595(6.6)	– 663(6.9)
Severity over all diagnoses (67)						
Elective only versus:						
Elective and mandatory	– 5.2(17.0)	– 5.1(15.9)	– 5.2(16.6)	– 65(.9)	– 51(.6)	– 65(.9)
Mandatory only	– 4.0(12.3)	– 4.0(11.6)	– 3.9(12.1)	– 318(3.8)	– 331(3.5)	– 319(3.9)
Number of diagnoses (45)						
1 versus:						
2	– 1.8(5.2)	– 1.8(5.2)	– 1.9(5.0)	32(.3)	53(.5)	26(.3)
3	– 4.2(10.0)	– 4.1(9.6)	– 4.2(9.8)	63(.8)	63(.7)	60(.8)
4 or more	– 7.8(19.5)	– 7.8(20.6)	– 7.7(19.8)	– 230(2.3)	– 257(2.2)	– 239(2.3)
Level of use						
Physician visits per person (59)						
1 versus:						
2 to 4	– 1.7(8.7)	– 1.8(83.1)	– 1.7(86.2)	—	—	—
5 or more	–11.1(40.2)	–11.1(39.5)	–11.1(40.1)	—	—	—
5 to 13	– 7.0(78.3)	– 7.0(72.8)	– 7.0(77.9)	—	—	—
14 or more	–23.8(30.9)	–23.8(29.3)	–23.7(29.8)	—	—	—
Expenditures per admission (23)						
$0 to 300 versus:						
301 to 600	—	—	—	– 242(38.1)	– 242(36.2)	– 242(38.2)
601 or more	—	—	—	– 1,341(12.2)	– 1,371(12.2)	– 1,340(12.0)
601 to 1200	—	—	—	– 657(36.2)	– 659(35.3)	– 657(37.1)
1201 or more	—	—	—	$– 2,585(10.4)	$– 2,616(11.1)	$– 2,586(10.1)

[a]Numbers in parentheses in the body of the table are the ratios of the differences to the estimated standard errors of the differences.

Table 10.1. Comparison of Three Alternative Estimates for Various Sample Allocation Plans

(k) Subsample Proportion	(n) Size of Total Social Survey Sample	(kn) Size of Validation Response Sample	TSE ($\bar{y}_s$) Substitution Estimate	TSE ($\bar{y}_r$) Two-Phase Ratio Adjustment Estimate	TSE ($\bar{y}_{rg}$) Two-Phase Regression Adjustment
.00000	1,000	0	125.00	—[a]	—[a]
.04444	900	40	123.33	149.90	141.25
.10000	800	80	121.25	88.77	84.69
.17143	700	120	118.57	71.01	68.51
.26667	600	160	115.00	65.10	63.44
.40000	500	200	110.00	65.34	64.25
.60000	400	240	102.50	71.02	70.42
.93333	300	280	90.00	84.55	84.46
1.00000	285	285	87.72	87.72	87.72

[a]Not applicable.

Table 10.2. Social Survey and Verification Means and Variances and their Covariances and Correlations for Hospital Expenditures and Physician Visits

	Hospital Expenditures	*Physician Visits*
Means		
Social survey	632.8	5.16
Verification	649.4	4.08
Per-element variances		
Social survey	1,095,014.0	53.39
Verification	1,045,706.0	31.25
Covariance ($S_{ss,v}$)	1,000,741.6	25.12
Correlation ($\rho_{ss,v}$)	.9352	.6150

Table 11.1. Effect of Bias Adjustment on Estimates for Groups of Policy Interest Relative to Estimates for Comparison Groups for Selected Health Care Characteristics

Characteristic	*Effect of Bias Adjustment on Group of Policy Interest Relative to Comparison Group*[a]					
	65 and over vs. Under 65[b]	*Poor vs. Nonpoor*	*Nonwhite vs. White*	*Farm vs. Suburban*	*Central City vs. Suburban*	*Poor Health vs. Excellent Health*
Inpatient Care						
Days per admission (33)	None	None	None	None	None	None
Expenditures per admission (23)	None	None	None	None	None	None
Out-of-pocket expenditures per admission (53)	Higher	None	None	None	None	—
Accuracy of hospital conditions	Higher	Higher	Higher	None	Higher	Higher
Ambulatory Care						
Visits per person seeing physician (59)	Lower	Lower	Lower	None	None	Lower
Office visit expenditures (24)	None	None	None	None	Lower	Lower
OPD/ER expenditures (14)	Lower	Higher	Higher	Higher	Higher	Higher
Accuracy of physician visit conditions	None	Higher	Higher	Higher	None	None
Health Insurance						
Health insurance premiums (61)	Higher	None	None	None	None	—
Hospital coverage (40)	Higher	Higher	None	None	Higher	—
Physician visit coverage (42)	Higher	Higher	Lower	Higher	Higher	—
Major medical coverage	Lower	Lower	None	Lower	None	—

[a] If the change in the difference between the policy group and the comparison group resulting from the bias adjustment exceeds the standard error of the difference between the groups, the effect is labeled: (1) "higher," if the adjusted estimate of the policy group divided by the adjusted estimate of the comparison group is greater than the unadjusted estimate of the policy group divided by the unadjusted estimate of the comparison group, (2) "lower," if the adjusted estimate of the policy group divided by the adjusted estimate of the comparison group is less than the unadjusted estimate of the policy group divided by the unadjusted estimate of the comparison group, or (3) "none," if the change is less than the standard error of the difference.

[b] Relevant variable is (6) for Accuracy of Hospital and Physician Visit Conditions.

Table 11.2. Proportional Bias and Proportion Overestimated and Underestimated in Social Survey for Selected Health Care Characteristics

Characteristic	*Proportional Bias*[a]	*Proportion of Observations in Social Survey*	
		Under-estimated	*Over-estimated*
Inpatient Care			
Days per admission (33)	102%	18%	42%
Expenditures per admission (23)	98	38	40
Out-of-pocket expenditures per admission (53)	149	11	30
Reporting of hospital conditions[e]	80[b]	38[c]	18[d]
Ambulatory Care			
Visits per person seeing physician (59)	116	28	37
Office visit expenditures (24)	108	40	41
OPD/ER expenditures (14)	76	31	46
Reporting of physician visit conditions[f]	89[b]	48[c]	37[d]
Health Insurance			
Health insurance premiums (61)	118	33	50
Hospital coverage (40)	105	4	8
Physician visit coverage (42)	110	8	18
Major medical coverage (41)	118	7	25

[a]Social Survey estimate as a percentage of adjusted estimate for those observations for which verification information was available.

[b]Social Survey diagnoses/matched social survey diagnoses plus unmatched verification diagnoses.

[c]Unmatched verification diagnoses/matched social survey diagnoses plus unmatched verification diagnoses.

[d]Unmatched social survey diagnoses/matched social survey diagnoses plus unmatched verification diagnoses.

[e]Universe of conditions as defined in Table 5.4.

[f]Universe of conditions as defined in Table 5.2.

APPENDIX A

Calculation of Biases

Judith Kasper

I. *Hospital admissions per 100 person-years (Chapter Two)*

The following discussion also applies to the following dependent variables: physician visit condition rates, hospitalized condition rates, and surgical procedure rates; see Chapter Five.

Notation	*Hospital Admissions*	
	Social Survey	*Verification*
a	Reported	Reported
b	Reported	Rejected
c	Reported	No verification obtained
d	No report	Reported

Note: I wish to thank Martha J. Banks for her helpful advice on organizing this section.

All admissions reported in the social survey is given by:

$$SS = a + b + c$$

A revised estimate of the number of admissions can be calculated by:

$$SS_R = a + d + \left[c \left(\frac{a}{a + b} \right) \right]$$

The adjustment to c, the number of admissions reported in the social survey for which no verification information was obtained, assumes that the same rate of rejection would have been observed as was the case for reported admissions for which we obtained verification data.

The estimate of field bias is the difference between the social survey estimate and the revised estimate. In addition, the estimate of field bias is converted to admissions per 100 person-years in the formula below:

$$\text{Field bias} = \frac{SS_R \times 12 \times 100}{N} - \frac{SS \times 12 \times 100}{N}$$

where

N = months all persons were in the social survey (persons who died or were institutionalized during the year were in the survey less than twelve months).

No processing bias for admissions can be calculated. Persons either reported an admission or did not; no imputation was involved.

Example: Calculation of field bias for hospital admissions per 100 person-years for rural farm residents (see Table 2.2).

Number of Admissions	*Hospital Admissions*	
	Social Survey	*Verification*
473.1	Reported	Reported
55.6	Reported	Rejected
26.5	Reported	No verification obtained
21.5	No report	Reported

$$SS = 473.1 + 55.6 + 26.5$$
$$= 555.2$$
$$SS_R = 473.1 + 21.5 + \left[26.5\left(\frac{473.1}{473.1 + 55.6}\right)\right]$$
$$= 518.4$$
$$\text{Field bias} = \frac{518.4 \times 12 \times 100}{45{,}754.8} - \frac{555.2 \times 12 \times 100}{45{,}754.8}$$

where

$N = 45{,}754.8$

Field bias $= -1.0$

II. *Hospital expenditures per admission (Chapter Three)*
The following discussion applies to several other dependent variables as well, including hospital days per admission (Chapter Two), insurance premiums per policy (Chapter Four), insurance coverage per policy (Chapter Four), and source of payment per admission (Chapter Four).

Hospital Expenditures per Admission

Admission	*Notation*	*Social Survey*		*Verification*	
		Expenditures	*Notation*	*Expenditures*	*Notation*
Admissions reported in social survey and verification	A	Reported	a_s	Reported	a_v
		Reported	b_s	Imputed	*
		Imputed	c_s	Reported	c_v
		Imputed	d_s	Imputed	*
Admissions reported in social survey, rejected in verification	B	Reported	e_s	Rejected	—
		Imputed	f_s	Rejected	—
Admissions reported in social survey, no verification obtained	C	Reported	g_s	No verification obtained	—
		Imputed	h_s	No verification obtained	—
No admission reported in social survey, reported in verification	D	No report	—	Reported	i_v

Note: * indicates that for purposes of bias calculations these were treated as no verification obtained.

The social survey estimate of hospital expenditures is given by all expenditures reported or imputed in the social survey, that is:

$$SS = a_s + b_s + c_s + d_s + e_s + f_s + g_s + h_s$$

An estimate of hospital expenditures revised for reporting error can be calculated by:

$$SS_R = SS + (i_v) - (e_s + f_s) + (a_v - a_s) + [(b_s + g_s)(a_v/a_s) - (b_s + g_s)]$$

The revised estimate is obtained by adding discovered expenditures (i_v), subtracting rejected expenditures (e_s, f_s), adding the difference between verification and social survey reports of expenditures $(a_v - a_s)$, and adjusting social survey data for which no verification was obtained (g_s) or verification data were imputed (b_s). Imputed expenditures that were rejected in verification (f_s) are treated as reporting error rather than imputation error because expenditures for an admission would not have been imputed if the respondent had not reported an admission. For expenditures reported in the social survey for which no verification was obtained or verification data were imputed, the adjustment factor (a_v/a_s) is the ratio of the verification and social survey reports of expenditures when reported data were obtained from both sources. Rejected and discovered data were not included in this adjustment term because the greater likelihood of rejecting data in our verification procedure was considered a potential source of bias.

The estimate of field bias per admission is calculated as follows:

$$\text{Field bias} = \frac{SS_R}{A + C + D} - \frac{SS}{A + B + C}$$

$$\begin{aligned} SS &= 158{,}391.6 + 17{,}484.1 + 216{,}134.6 + 58{,}132.4 + \\ &\quad 7{,}428.2 + 44{,}463.9 + 17{,}064.6 + 15{,}661.1 \\ &= 534{,}760.5 \\ SS_R &= 534{,}760.5 + 35{,}751.8 - (7{,}428.2 + 44{,}463.9) + \\ &\quad (164{,}620.4 - 158{,}391.6) + [(17{,}484.1 + 17{,}064.6)\left(\frac{164{,}620.4}{158{,}391.6}\right) - \\ &\quad (17{,}484.1 + 17{,}064.6)] \\ &= 526{,}174.8 \end{aligned}$$

$$\text{Field bias} = \frac{526{,}174.8}{632 + 49 + 49} - \frac{534{,}760.5}{632 + 97 + 49}$$
$$= 33.5$$

$$SS_I = 534{,}760.5 + (180{,}064.5 - 216{,}134.6) +$$
$$[(15{,}661.1 + 58{,}132.4)\left(\frac{180{,}064.5}{216{,}134.6}\right) - (15{,}661.1 + 58{,}132.4)]$$
$$= 486{,}145.5$$

$$\text{Processing bias} = \frac{486{,}145.5}{632 + 97 + 49} - \frac{534{,}760.5}{632 + 97 + 49}$$
$$= -62.5$$

A revised estimate of hospital expenditures adjusting for imputation error can be calculated by:

$$SS_I = SS + ({}^c v - {}^c s) + [(h_s + d_s)\ (c_v/c_s) - (h_s + d_s)]$$

For expenditures imputed in the social survey for which no verification was obtained or verification data were imputed, the adjustment factor $({}^c v/{}^c s)$ is the ratio of verification and social survey expenditures for admissions when the former was reported and the latter imputed.

The estimate of processing bias per admission is calculated as follows:

$$\text{Processing bias} = \frac{SS_I}{A + B + C} - \frac{SS}{A + B + C}$$

Example: Calculation of field bias and processing bias for hospital expenditures per admission for nonwhites (see Table 3.2).

		Hospital Expenditures per Admission			
		Social Survey		Verification	
Admission	Number	Expenditures	Dollars	Expenditures	Dollars
Admissions reported in social survey and verification	632	Reported	158,391.6	Reported	164,620.4
		Reported	17,484.1	Imputed	*
		Imputed	216,134.6	Reported	180,064.5
		Imputed	58,132.4	Imputed	*
Admissions reported in social survey, rejected in verification	97	Reported	7,428.2	Rejected	—
		Imputed	44,463.9	Rejected	—
Admissions reported in social survey no verification obtained	49	Reported	17,064.6	No verification obtained	—
		Imputed	15,661.1	No verification obtained	—
No admission reported in social survey, reported in verification	49	No report	—	Reported	35,751.8

Note: * indicates that for purposes of bias calculations these were treated as no verification obtained.

III. *Percent seeing a physician (Chapter Two)*

	Physician Contact	
Notation	Social Survey	Verification
a	Reported seeing physician	Reported seeing patient
b	Reported seeing physician	Rejected by all physicians
c	Reported seeing physician	No verification obtained
d	Reported no physician	Reported seeing patient

All persons who reported seeing a physician in the social survey is given by:

$$SS = a + b + c$$

A revised estimate of the number of persons seeing a physician can be calculated by:

$$SS_R = a + d + \left[c\left(\frac{a}{a+b}\right)\right]$$

The adjustment to c, persons who reported seeing a physician but for whom no verification data were obtained from any of the physicians they reported seeing, assumes the same rate of rejection would have been observed as was the case for persons for whom verification data were obtained.

The estimate of field bias for the proportion of the population seeing a physician is the difference between the social survey estimate and the revised estimate,

$$\text{Field bias} = \frac{SS_R}{N} - \frac{SS}{N}$$

where

N = all persons in the social survey

No processing bias for physician contact can be calculated. Persons either reported seeing a physician during the year or did not; no imputation was involved.

Example: Calculation of field bias for physician contact for persons who perceived their health as excellent (see Table 2.6).

Number of Persons	*Physician Contact*	
	Social Survey	*Verification*
7,828.6	Reported seeing physician	Reported seeing patient
1,200.5	Reported seeing physician	Rejected by all physicians
4,244.0	Reported seeing physician	No verification obtained
45.4	Reported no physician	Reported seeing patient

$$SS = 7{,}828.6 + 1{,}200.5 + 4{,}244$$
$$= 13{,}273.1$$

$$SS_R = 7{,}828.6 + 45.4 + 4244\left(\frac{7{,}828.6}{7{,}828.6 + 1200.5}\right)$$
$$= 11{,}553.7$$

$$\text{Field bias} = \frac{11{,}553.7}{21{,}452} - \frac{13{,}270.1}{21{,}452}$$

where

$N = 21{,}452$
Field bias $= -8.0\%$

IV. *Expenditures per person for emergency room, outpatient, or clinic visits (Chapter Three)*

The following discussion applies to several other dependent variables as well, including physician visits per person with visits (Chapter Two) and expenditures per person for office visits (Chapter Three).

Expenditures per Person for Emergency Room, Outpatient, or Clinic Visits

Number of Persons	*Social Survey*		*Verification*	
	Expenditures	*Notation*	*Expenditures*	*Notation*
A	Reported only	a_s	Reported only	a_v
B	Reported only	b_s	All rejected	—
C	Reported only	c_s	Reported and imputed Imputed only Partial or no verification	*
D	No report	—	Reported and/ or imputed	d_v
E	Any imputation	e_s	Reported only	e_v
F	Any imputation	f_s	All rejected	—
G	Any imputation	g_s	Reported and imputed Imputed only Partial or no verification	*

Note: * indicates that for purposes of bias calculations these were treated as no verification obtained.

Comparison between social survey and verification data was made on a per person basis for expenditures for emergency room, outpatient, or clinic visits. Therefore, it is not possible to calculate a bias in expenditures per physician visit in the same way a bias in expenditures per hospital admission was calculated. To calculate a field and processing bias, we classified persons according to whether they only had reported data in the social survey or whether any imputation of data occurred (whether data were also reported). In contrast to previous bias calculations, field bias was calculated using that portion of the social survey data for respondents with reported data only (*SSR*). Similarly, processing bias was calculated using only that portion of the social survey data for respondents with any imputed data (*SSI*). The relationship of *SSR* and *SSI* is:

$$SS = SSR + SSI$$

All emergency room, outpatient or clinic visits expenditures *reported* in the social survey is given by:

$$SSR = a_s + b_s + c_s + f_s$$

A revised estimate of reported social survey expenditures can be calculated by:

$$SSR' = [SSR - (b_s + f_s)]\,(a_v/a_s) + d_v$$

An individual may have several kinds of verification outcomes; for example, one doctor agrees with the respondent's report but no verification is obtained from another physician. In calculating the field bias, expenditures were subtracted from the social survey total if reported expenditure data for that person were rejected (b_s, f_s). Imputed expenditures that were rejected in verification (f_s) are treated as reporting error rather than imputation error because expenditures for these visits would not have been imputed if the respondent had not reported these visits. This number is then adjusted by the error rate for persons who had only reported expenditure data in both the social survey and verification (a_v/a_s). Finally, expenditures were added for persons who were discovered to have had expenditures for emergency room, outpatient, or clinic visits that they did not report (d_v).

The difference between the social survey estimate of reported expenditures for emergency room, outpatient, or clinic visits and the revised social survey estimate is:

$$SSR'' = \frac{SSR'}{A + C + D} - \frac{SSR}{A + B + C + F}$$

To produce an average field bias per person with expenditures in the survey, we weighted this difference (SSR'') by the proportion of persons with only reported social survey expenditures (SSR) to all persons with social survey expenditures (SS).

$$\text{Field bias} = SSR'' \left(\frac{A + B + C + F}{A + B + C + E + F + G} \right)$$

All emergency room, outpatient, or clinic visit expenditures for persons with any *imputed* expenditures is given by:

$$SSI = e_s + g_s$$

A revised estimate of expenditures for persons with any imputed survey expenditures can be calculated by:

$$SSI' = SSI\ (e_v/e_s)$$

All social survey data for persons with any imputed data are adjusted by the error rate for persons with any imputed expenditures in the social survey and only reported expenditures in the verification (e_v/e_s).

The difference between the social survey estimate of expenditures for persons with any imputed data and the revised social survey estimate is:

$$SSI'' = \frac{SSI'}{E + G} - \frac{SSI}{E + G}$$

To produce an average processing bias per person in the survey, we weighted this difference (SSI'') by the proportion of persons with any imputed social survey expenditures (SSI) to all persons with social survey expenditures (SS).

$$\text{Processing bias} = SSI'' \; \frac{E + G}{A + B + C + E + F + G}$$

Example: Calculation of field bias and processing bias for emergency room, outpatient, or clinic visit expenditures for persons with four or more diagnoses (Table 3.6).

Expenditures per Person for Emergency Room, Outpatient, or Clinic Visits

Number of Persons	Social Survey		Verification	
	Expenditures	Dollars	Expenditures	Dollars
233	Reported only	15,688.8	Reported only	9,115.6
38	Reported only	1,067.8	All rejected	—
420	Reported only	22,363.5	Reported and imputed Imputed only Partial or no verification	*
265	No report	—	Reported and/ or imputed	12,646.6
61	Any imputation	4,319.9	Reported only	8,217.4
346	Any imputation	9,707.7	All rejected	—
439	Any imputation	44,652.6	Reported and imputed Imputed only Partial or no verification	*

Note: * indicates that for purposes of bias calculation these were treated as no verification obtained.

$$SSR = 15{,}688.8 + 1{,}067.8 + 22{,}363.5 + 9{,}707.7$$
$$= 48{,}827.8$$

$$SSR' = \left[48{,}827.8 - (1{,}067.8 + 9{,}707.7)\right]\left[\frac{9{,}115.6}{15{,}688.8}\right] + 12{,}646.6$$
$$= 34{,}716.99$$

$$SSR'' = \frac{34{,}716.9}{233 + 420 + 265} - \frac{48{,}827.8}{233 + 38 + 420 + 346}$$
$$= -9.3$$

$$\text{Field bias} = -9.3\left(\frac{1{,}037}{1{,}537}\right)$$
$$= -6.2$$

$$SSI = 4{,}319.9 + 44{,}652.6$$
$$= 48{,}972.5$$

$$SSI' = 48{,}972.5\left(\frac{8{,}219.4}{4{,}319.9}\right)$$

$$= 93{,}156.5$$

$$SSI'' = \frac{93{,}156.5}{500} - \frac{48{,}972.5}{500}$$

$$= 88.4$$

$$\text{Processing bias} = 88.4\left(\frac{500}{1{,}537}\right)$$

$$= 28.8$$

APPENDIX B

Example of Adjusting Factors When One of Them Lies Outside of a Desired Range

* * * * * * * * * * * * * * * * * *

Table B.1.

Category	Percent in Category		Post-Stratification Factor
	Control Data	*Sample Data*	
	A. Original Status		
1	14.2778	15.3785	.9284
2	12.9711	6.1892	2.0958
3	16.6519	18.0553	.9223
.	.	.	.
.	.	.	.
.	.	.	.
Total	100.0000	100.0000	
	B. Status after adjustment[a]		
1	14.2778	15.3785	.9284
2	12.9711 − .5927 = 12.3784	6.1892	2.0000
3	16.6519 + .5927 = 17.2446	18.0553	.9551
.	.	.	.
.	.	.	.
.	.	.	.
Total	100.0000	100.0000	

[a]If 2.0000 has been chosen as the maximum factor to be used, then category 2 must be adjusted. If we decide that category 3 is most similar to category 2 and can be adjusted in a compensating way, then the adjusted worksheet is as presented in part B.

APPENDIX C

Definitions of Variables and Variable Frequencies

1–5. *Accuracy*

Accuracy was measured by the proportion of social survey to verification information:

$$\frac{\text{social survey} - \text{verification}}{\text{social survey} + \text{verification}}$$

Persons whose reporting accuracy scores were within 25 percent of the verification information were considered "accurate" reporters. A person underreported if the verification exceeded the social survey information by more than 25 percent. A person overreported if the social survey exceeded the verification by more than 25 percent.

1. *Reporting accuracy for hospital admissions*

   Accuracy with which the number of hospital admissions for each member of the household was reported. (See definition 31.)

2. *Reporting accuracy for hospital days*
   Accuracy with which the number of days each member of the household spent in a hospital was reported. (See definition 35.)
3. *Reporting accuracy for hospital expenditures*
   Accuracy with which the hospital expenditures for each member in the household were reported. (See definition 19.)
4. *Reporting accuracy for physician visits*
   Accuracy with which number of physician visits for each member in the household was reported. (See definition 59.)
5. *Reporting accuracy for office visit expenditures*
   Accuracy with which expenditures for physician office visits for each member in the household were reported. (See definition 24.)

6. *Age*
   Age as of the last day of the study year, that is, December 31, 1970.

7. *Age of oldest family member*
   Age of the oldest member of a sample family as of the last day of the study year, that is, December 31, 1970. Families are grouped into those with all family members less than sixty-five years old and those with the oldest member sixty-five years or older.

8. *Age of physician*
   Age of physician as listed in the 1969 AMA Directory. For physicians not listed in the directory, no age is coded.

9. *Board certification*
   For each physician mentioned in the social survey who was listed in the 1969 AMA Directory, whether or not he or she was board certified (that is, whether the physician had passed a specialty board examination following completion of a residency).

10. *Clinic cooperation*
Whether clinic complied with our request for verification of social survey information. A clinic may be counted more than once if more than one respondent named it. It was weighted each time by the weight of the patient who named the clinic as providing care.

11. *Cooperation on calls by interviewer*
Each time the interviewer visited a residence to obtain information, the respondent was given a cooperation code. These codes were summarized for each family as follows:
a. The outcome of the call was successful (that is, interviewer did not report that the respondent refused to answer the door, answer questions, or supply additional information or that the respondent requested more information about the survey before responding) and the respondent gave written permission for verification of his or her responses.
b. Outcome was successful but only oral permission for verification was given.
c. Outcome was successful but respondent refused to sign the permission form or give oral permission for verification.
d. Respondent requested more information about the survey before responding to any questions.
e. Respondent never refused to cooperate, but one outcome was unsuccessful (for example, an appointment was broken or respondent did not answer the door and interviewer suspected someone was at home).
f. Respondent never refused to cooperate but more than one outcome was unsuccessful.
g. At some point respondent refused additional information or further cooperation with the survey.
Persons with codes of a–d are considered "cooperative"; those with codes of e, f, or g are "uncooperative."

12. *Education*
Education for each person is based on the response to the question: "What is the highest grade or year (PERSON) has completed in school?" Interviewers were instructed not to

include trade schools, business colleges, correspondence courses, and the like in the calculation of years of formal education.

13. *Enrollment status of policy*
This is the method through which individuals and families purchase health insurance. The two categories used in this study are:
a. Work group—Enrollment through an organization, usually an employer group, covered by one general health insurance policy. In this category subdivisions are made for cases in which the employer pays (1) all, (2) some, or (3) none of the premium for the policy.
b. Not work group—Enrollment by direct purchase of a single policy covering individual or family from an insurer or insurance agent. This includes conversions from a group.

14. *Expenditures, emergency room, outpatient, or clinic*
Charges for visits per person to an organized outpatient department of a hospital when the patient was seen by salaried house staff. Also included are charges for visits to the emergency room of a hospital as distinct from a regular or drop-in clinic appointment and visits to any other site, including employee and student health facilities, board of health clinics not connected with a hospital, health maintenance organizations (HMOs), and certain privately maintained union and fraternal clinics not connected with hospitals. Charges for visits to fee-for-service physicians at these sites are included in office visits because the physician is then considered to maintain a private office at the site in question. Included here are charges for care by the practitioner and for special tests and x rays, treatments, drugs, and medications when the physician bills the patient for such services without necessarily performing them personally.

15. *Expenditures, family dental*
Charges by the dentist for his or her services and those of auxiliary personnel such as a dental hygienist and for dental

appliances. Includes also expenditures to cover charges to the dentist by dental laboratories and dental manufacturers for work done for the patient. Excludes in-hospital extractions and other surgery.

16. *Expenditures, family drug*

This is the sum of the following two components of drug expenditures per family:

a. Prescribed drugs—Charges for drugs and medicines prescribed at some time by a physician, other medical practitioner, or dentist and purchased by the consumer directly from a pharmacy or elsewhere. It excludes drugs administered by the physician or dentist and charged for on his or her bill as well as drugs received from a hospital and included in the hospital bill.

b. Nonprescribed medicines and drugs—Charges for all medicines, tonics, vitamins, aspirin, lotions, and so forth that are "over-the-counter" purchases rather than being prescribed.

17. *Expenditures, family health*

Total family expenditures, both free and nonfree, are included.

Free expenditures are expenditures for medical goods and services provided to the family at no direct cost or at substantially reduced rates and without benefits being provided by any type of health insurance plan. Included are:

a. Medicaid, welfare, free institutions—This includes expenditures for medical care made by public aid departments on behalf of their recipients and includes as well expenditures for those not on welfare but with incomes sufficiently low to qualify them for payment of some or all of their medical bills (Medicaid). Eligibility requirements vary widely from state to state for both of these programs. The main distinction between them is that welfare recipients receive a monthly support check and Medicaid recipients do not. Also included here is the estimated monetary value of care provided by locally tax-supported hospitals and institutions, including city and county hospitals, state mental and tuberculosis hospitals, and board of health clinics.

b. Other free expenditures—This variable includes all expenditures that have no monetary value attached to them and are not included in Medicaid, welfare, and free institutions above. This care, although free, generally does not have stringent income requirements for eligibility and thus is treated separately. Included here is care provided by Veteran's Administration hospitals, care provided to armed forces members, workmen's compensation, care provided by certain governmental agencies such as federally funded crippled children's programs, and care paid for by liability insurance carried by a nonfamily member.

Nonfree expenditures are generally considered to be expenditures for care that the consumer pays for either directly or through some third party to which he or she pays premiums or has premiums paid. Its four subdivisions are:

a. Voluntary insurance—This is health insurance carried either directly by the patient or by his or her employer. Insurance that pays for accidental injuries only is excluded.

b. Medicare—This insurance is funded through Social Security payments for people sixty-five and over. Part A, which covers inpatient hospital care, requires no additional payment. Part B, which covers physician care, requires a monthly premium paid by those who carry it. Medicare is considered nonfree care in this report because there are no income restrictions on eligibility and because in large part it replaces voluntary health insurance carried by people sixty-five and over.

c. Out-of-pocket expenditures—These are expenditures made directly either at time of service or to pay off medical bills incurred previously during the year. Out-of-pocket expenditures also include money owed but not yet paid for care received in 1970.

d. Other nonfree expenditures—This variable consists of two items that are not voluntary health insurance. One is

CHAMPUS, insurance provided by the armed forces for dependents of members. The other is payments made by accident and liability insurance carried by some member of the patient's family.

18. *Expenditures, family hospital*
Sum of all charges incurred by members of the family for inpatient admissions. (See definition 19 for further explanation.)

19. *Expenditures, hospital*
All charges incurred in connection with an inpatient admission per person. They include room and board charges, laboratory fees, drugs, x rays, operating and delivery room fees, and the usual "extras." They include charges for pathologist, radiologist, and anesthesiologist if these are included in the hospital bill rather than being billed privately. Charges for special-duty nursing are always excluded.

20. *Expenditures, hospital, for stay of operation*
All charges incurred in connection with an inpatient admission during which an operation occurred. (These charges cover services described in definition 19.)

21. *Expenditures, out-of-pocket hospital*
The sum of out-of-pocket expenditures the individual made directly either at time of service or to pay off hospital charges incurred previously during the year. Out-of-pocket expenditures also include money owed but not yet paid for care received in 1970. (See definition 19 for an explanation of services covered in these expenditures.)

22. *Expenditures, out-of-pocket physician visit*
These are individual expenditures for physician visits made directly either at time of service or to pay off medical bills incurred previously during the year. Out-of-pocket expenditures also include money owed but not yet paid for care received in 1970. Includes charges for care by the practitioner

and for special tests, x rays, treatments, drugs, and medications when the physician bills the patient for such services without necessarily performing them personally.

23. *Expenditures per admission*

    Charges per admission incurred in connection with an inpatient admission. (See definition 19 for services covered.)

24. *Expenditures, physician office visit*

    Charges per person for care delivered at a fee-for-service physician's private office, including those offices located in clinics or clinic buildings. Also includes visits to prepaid group practice physicians when these were designated "office." Includes charges for care by the practitioner and for special tests, x rays, treatments, drugs, and medications when the physician bills the patient for such services without necessarily performing them personally.

25. *Expenditures, prescribed drugs*

    Total charges for an individual for drugs and medicines prescribed at some time by the physician, other medical practitioner, or dentist and purchased by the consumer directly from a pharmacy or elsewhere. It excludes drugs administered by the physician or dentist and charged for on the bill as well as drugs received from a hospital and included in the hospital bill.

26. *Family income*

    Total family money income before taxes for the survey year. Income from wages, salaries, own business or farm, professional work or trade, pensions, rents, welfare agencies, unemployment compensation, alimony, child support, regular contributions from friends or relatives, dividends, interest, and similar sources are included. Income in kind—the value of free rent or noncash benefits—is excluded. Data were obtained through questions covering the earned and unearned income of each individual fourteen or older in the family. Total family income is the sum of these components.

A "poor" family is defined as a family with an income less than the following amounts for a given family size for 1970:

Family Size	*Family Income*
1	$2,600
2	3,700
3	4,500
4	5,700
5	6,600
6	7,500
7 or more	9,100

These were determined using Bureau of Labor Statistics figures for 1969 updated for 1970 and increased by 25 percent.

27. *Family size*

A family is defined as one person or a group of persons living together and related to each other by blood, marriage, or adoption. More informal arrangements are also recognized as long as the respondents consider themselves a family (that is, common-law marriage, foster children). When there were two related married couples living in a single dwelling unit, each married couple and its unmarried children were considered a separate family. Any person unrelated to anyone else in the dwelling unit, such as a roommate, boarder, or live-in housekeeper, was considered a separate family.

A "small" family is defined as having one or two members. A "medium-sized" family has three to five members. A "large" family includes six or more members.

28. *Hospital admissions, family*

Total number of hospital admissions claimed for the family. (See 30 for definition of hospital admission.)

29. *Hospital admissions per condition*

Number of overnight stays in or surgeries performed in hospitals for a particular hospitalized condition. (Hospital admissions as defined in 31.)

30. *Hospital admissions per 100 person-years*
    Hospital admission is defined as an overnight stay in or surgery performed in hospitals. (See definition 31 for a full explanation.)

    Person-years are computed by summing the total months by sample members in the population universe during the survey year and dividing this sum by 12. The purpose of this base is to adjust for sample members who were not in the population the entire survey year, such as those who died, were institutionalized, or were born during the year.

31. *Hospital admissions per person* (overnight stays or surgery performed in a hospital)
    The total admissions reported for each individual. Included are admissions to long-term, mental, and tuberculosis hospitals. The delivery admission for an obstetrical case is counted as one admission for the mother and none for the infant. If the infant stays in the hospital after the mother goes home or if the infant is readmitted after being discharged, a separate admission is counted for the infant. Stays are included even if all the days of the stay do not fall within the survey year.

32. *Hospital days for stay of operation*
    Total number of days' duration for hospital stay during which the operation was reported in the social survey.

33. *Hospital days per admission*
    Total number of days' duration for each hospital stay during the survey year, regardless of when the stay began or ended.

34. *Hospital days per condition*
    Sum of days spent in hospitals over all admissions involving a reported condition.

35. *Hospital days per person*
    For those who reported an admission, total number of days spent in hospitals during the survey year, regardless of when the admission began or ended.

36. *Hospital expenditures were reported*
Respondent reported family hospital expenditures (see definitions 18 and 19) for the survey year in the social survey.

37. *Income was reported*
Respondent agreed to divulge family income (as described in definition 26) in the social survey.

38–43. *Insurance coverage*

38. *Dental coverage*
Dental coverage pays for all or part of the cost of outpatient dentistry. It can be either a separate plan limited to dentistry or part of a larger plan, such as a prepaid group or major medical policy.

Policies are excluded from this analysis if the sample family did not provide information as to whether the policy included dental coverage.

39. *Drug coverage*
Drug coverage pays for all or part of the cost of prescribed drugs other than those provided to hospital inpatients. Major medical insurance that pays for prescribed drugs only for a specified illness, only when a deductible has been met, or both is *not* considered drug insurance in this definition.

Policies are excluded from this analysis if the sample family did not provide information as to whether the policy included drug coverage.

40. *Hospital coverage*
Hospital coverage pays all or part of the inpatient hospital bill for the insured family. The hospital bill includes only the bill submitted by the hospital itself, not that submitted by the doctor who treated the patient in the hospital nor those submitted by private-duty nurses. Flat rate insurance that pays a given amount for every day

the person is hospitalized *is* considered hospitalization insurance. Major medical insurance is considered hospital insurance *only* if it includes supplementary basic coverage for hospitalizations, that is, the coinsurance, deductible, or both for the major medical policy as a whole is waived or reduced for inpatient hospital expenses. Policies are excluded from this analysis if the sample family did not provide information as to whether the policy included hospital coverage.

41. *Major medical coverage*
Major medical coverage pays for the heavy medical expenses resulting from catastrophic prolonged illness. It typically includes a deductible (for example, $100–$250), coinsurance (for example, 20–25 percent), and a high maximum payment (for example, $100,000 or more). Within these limits it covers most out-of-hospital as well as in-hospital expenses associated with the particular illness.

Policies are excluded from this analysis if the sample family did not provide information as to whether the policy included major medical coverage.

42. *Physician visit coverage*
Physician visit coverage pays for all or part of the cost of seeing the physician in his or her office, the patient's home, or both. Major medical policies that pay a percentage of the cost for physician office visits after a deductible has been paid by the patient are *not* considered physician visit insurance in this analysis.

Policies are excluded from this analysis if the sample family did not provide information as to whether the policy included physician visit coverage.

43. *Surgical-medical coverage*
Surgical-medical coverage pays all or part of the surgeon's bill in or outside of the hospital. In addition, the

medical portion of the insurance pays for visits by a physician other than for surgery when the patient is hospitalized. Major medical insurance is considered surgical-medical insurance *only* if it includes supplementary basic surgical-medical coverage, that is, the coinsurance, deductible, or both for the major medical policy as a whole is waived or reduced for surgical-medical expenses. Policies are excluded from this analysis if the sample family did not provide information as to whether the policy included surgical-medical coverage.

44. *Number of calls to complete main questionnaire*
Number of times interviewer called or visited the household before the main questionnaire was completed.

45. *Number of diagnoses*
For each person, number of reasons given for visiting a physician. In cases where both social survey and verification information were collected, the figures given in verification were used. Otherwise the numbers reported in social survey were used.

46. *Number of hospitalized conditions*
Number of reasons for which sample person was hospitalized during the survey year.

47. *Number of physician visit conditions*
Number of unique conditions for which sample person visited a physician in 1970. Condition is defined as the reasons respondent gave for sample person's visits to a physician.

48. *Number of surgical procedures*
Number of surgical procedures (including cesarean but not normal deliveries) or settings of a dislocation or fracture performed on sample person during the survey year. Endoscopic procedures, suturing of wounds, and circumcision of newborn infants, often classified as surgical procedures, are not so classified in this study. A few exceptions were made when the

suturing was so extensive as to require an operating room or blood transfusions. Dentists performing inpatient procedures are included.

49. *Number of times verification requested*
Number of times a physician or clinic was named by a social survey respondent and we attempted to verify information about that respondent.

50. *Operations for hospitalized condition*
Number of surgical procedures performed on a hospital inpatient during all stays for a hospitalized condition. (Surgical procedure is discussed in definition 48.)

51–54. *Payments for hospital admissions*

51. *Insurance payments*
These are expenditures per admission paid by voluntary health insurance carried either directly by the patient or by his or her employer. Payments from policies covering only accidental injuries are excluded. (See definition 19 for an explanation of hospital services covered.)

52. *Medicare payments*
These are expenditures per admission paid by insurance funded through Social Security payments for people sixty-five and over. (Medicare is further explained in definition 17. Services covered are discussed in definition 19.)

53. *Out-of-pocket payments*
These are expenditures made directly either at time of service or to pay off medical bills incurred previously during the year. Out-of-pocket expenditures also include money owed but not yet paid for care received in 1970. (See definition 19 for explanation of hospital services covered.)

54. *Welfare payments*
This includes expenditures per admission for medical care covered by welfare. (Welfare is discussed in definition 17.)

55. *Perception of health*
Respondent was asked to evaluate each household member's state of health with the question: "Would you say (PERSON)'s health in general is excellent, good, fair, or poor?"

56. *Physician contact*
Whether respondent reported at least one physician visit during the survey year. Visits by a doctor to a hospital inpatient are included. Excluded are all persons not in the universe for all twelve months of the survey year.

57. *Physician cooperation*
Whether physician complied with our request for verification of social survey information. A physician may be counted more than once if more than one respondent named him or her. The physician was weighted each time by the weight of the patient who named him or her as providing care.

58. *Physician expenditures were reported*
Respondent reported family expenditures for physician visits (see definition 59) for the survey year in the social survey.

59. *Physician visits*
Sum of all visits per person related to hospitalized illness, other major nonhospitalized illness, pregnancy, other minor illness, and routine checkups, shots, tests, and ophthalmologist visits for the survey year. Includes seeing either a doctor or osteopath or his or her nurse or technician at the following sites: patient's home, doctor's office or private clinic, hospital outpatient service, and any other clinic such as a board of health clinic or neighborhood health center. Excluded are telephone calls and visits by a doctor to a hospital inpatient.

60. *Physician visits, family*

Sum of all visits for the family that were related to hospitalized illness, other major nonhospitalized illness, pregnancy, other minor illness, and routine checkups, shots, tests, and ophthalmologist visits for the survey year. (See definition 59.)

61. *Premium for health insurance policy*

The cost to the family for coverage during 1970 of a plan specifically designed to pay all or part of the health expenses of any sample person that was in effect all or part of the calendar year 1970. (The specific health services included for different types of coverage are defined under definitions 38–43.) Included in the analysis are plans provided by the following types of insurers: (1) Blue Cross-Blue Shield, (2) private insurance (for example, Aetna, Mutual of Omaha), and (3) independent plans (for example, Kaiser). Excluded from the analysis is all coverage of health expenses provided by Medicare, Medicaid, welfare, CHAMPUS, and all other third-party payers not listed in the definition of included insurers. (For detailed definitions of the plans included and excluded see Andersen, Lion, and Anderson, 1976, pp. 269–270, 277–278). Also excluded are cases where premium amount was unknown.

62. *Race*

Each family member is coded according to the race of the main respondent. The census definitions of white and nonwhite are used. People of Mexican or Spanish descent are coded as white. American Indians and Orientals are coded as nonwhite. When the distinction is made between white and black, persons of Mexican or Spanish descent are excluded from the white category and American Indians and Orientals from the black category.

63. *Regular source of care*

Whether respondent named a particular physician or clinic as individual's "regular source of care."

64. *Residence*

A classification of the residence of each person in the sample was done according to U.S. Census Bureau designation of the locality in which the residence is located plus the interviewer's description of the dwelling unit and locality. The following divisions were made:

a. Rural nonfarm—Residence in areas defined as rural by the census that are not described as "farms" by the interviewer.

b. Rural farm—Residence in areas defined as rural by the census that are described as "farms" by the interviewer.

c. SMSA central city—Residence in the urban part of a Standard Metropolitan Statistical Area (SMSA) according to the census that is also designated by the interviewer as "inside the largest city in the primary unit" (NORC's PSUs).

d. SMSA other urban—Defined as (c) except interviewer did not describe dwelling as "inside largest city."

e. Urban nonSMSA—All other urban residences.

65. *Response to verification request*

Answer of respondent to interviewer's request for permission to verify the social survey data by contacting physicians, hospitals, and insurance companies mentioned by the respondent.

66. *Severity of condition*

Each diagnosis or condition for which an individual reported being seen in a physician visit or being hospitalized was rated by a panel of five physicians according to whether the care generally provided was (1) for prevention or reassurance only, (2) for symptomatic relief, (3) for a condition for which a physician should be seen, or (4) for a condition for which a physician must be seen.

67. *Severity over all diagnoses* (based on verification data)

Each diagnosis or condition for which an individual was seen during 1970 was rated and given a classification. (Explained in definition 66.)

68. *Sex*
In those few cases where the interviewer did not specify the sex of a sample member, classification was made in the office on the basis of name or other information provided in the interview. If this was not possible, the interviewer, the respondent, or both were contacted again for the correct classification.

69. *Sex and employment status*
Each person was coded by sex and by whether he or she was employed full time.

70. *Specialty*
The specialty reported in the 1969 AMA Directory for listed physicians. Specialties were coded as general practice, general surgery, primary care specialties, and "other" specialties. Primary care specialties include pediatrics, internal medicine, and obstetrics-gynecology.

71. *Survey respondent*
The person (usually the head of the household or spouse) answering the survey questions. Interviewer selected the respondent with the following statement: "I'd like to talk to the person who knows the most about the health care and medical costs for your family during the last year." Two main respondents were permitted per household.

72. *Threat level*
Degree of threat or embarrassment assumed to be associated with a particular physician visit condition, a particular surgical procedure, or a particular hospitalized condition. Adapted from Cannell (NCHS, 1965a).

73. *Worry about health*
Respondent was asked, "Over the past year has (PERSON)'s health caused a great deal of worry, some worry, hardly any worry, or no worry at all?" for each member of the family.

Table C.1.

Variable Number	Description	Classes	Weighted N	Unweighted N
1 to 5	Accuracy			
1	Reporting accuracy for hospital admissions	1. Underreported	294	77[a]
		2. Overreported	429	122
		3. Accurately reported	5,395	1,128
		4. Not applicable	50,697	10,291
2	Reporting accuracy for hospital days	1. Underreported	331	90[a]
		2. Overreported	761	209
		3. Accurately reported	4,943	1,007
		4. Not applicable	50,780	10,312
3	Reporting accuracy for hospital expenditures	1. Underreported	368	104[a]
		2. Overreported	392	98
		3. Accurately reported	3,506	602
		4. Not applicable	52,549	10,814
4	Reporting accuracy for physician visits	1. Underreported	4,926	990[a]
		2. Overreported	7,016	1,471
		3. Accurately reported	13,625	2,390
		4. Not applicable	31,245	6,767
5	Reporting accuracy for office visit expenditures	1. Underreported	4,096	889[a]
		2. Overreported	7,016	1,155
		3. Accurately reported	9,377	1,518
		4. Not applicable	36,326	8,056
6	Age	a. 1. Less than 6 years	5,826	1,283[a]
		2. 6 to 17	14,591	3,166
		3. 18 to 34	12,907	2,419
		4. 35 to 54	12,476	2,226
		5. 55 to 64	5,280	1,019
		6. 65 or more	5,741	1,506
		b. 1. Less than 17 years	19,299	4,225[a]
		2. 17 to 40	17,545	3,326

		3. 41 to 64	14,234	2,562
		4. 65 or more	5,741	1,506
		c. 1. 17 to 25 years	7,972	1,558[a]
		2. 26 to 45	12,953	2,344
		3. 46 to 55	6,301	1,102
		4. 56 to 65	5,052	984
		5. 66 or more	5,242	1,406
		6. Excluded: less than 17 years	19,299	4,225
7	Age of oldest family member	1. Less than 65 years	48,996	9,451[a]
		2. 65 years or more	7,821	2,168
8	Age of physician	1. Less than 36 years	4,342	900[n]
		2. 36 to 45	17,196	2,909
		3. 46 to 55	16,189	2,965
		4. 56 to 65	10,660	1,915
		5. 66 years or more	2,806	539
		6. DK, NA, not in AMA Directory	2,970	494
9	Board certification	1. Yes	15,464	2,414[n]
		2. No	36,017	6,834
		3. DK, not in AMA Directory	2,675	474
10	Clinic cooperation	1. Cooperated	5,970	1,964[p]
		2. Did not cooperate	778	256
11	Cooperation on calls by interviewer	1. Cooperative	50,893	10,702[a]
		2. Uncooperative	5,923	917
12	Education	1. Less than 9 years	7,945	2,118[a]
		2. 9 to 11	6,620	1,473
		3. 12	11,013	1,885
		4. 13 years or more	9,508	1,405
		5. DK, NA, under 19 years of age	21,732	4,738
13	Enrollment status of policy	1. Work group, employer pays none	1,341	268[g]
		2. Work group, employer pays some	5,853	920
		3. Work group, employer pays all	6,173	860
		4. Not work group	7,169	1,515
		5. DK, NA, found in verification	665	138

Table C.1. (Continued)

Variable Number	*Description*	*Classes*	*Weighted N*	*Unweighted N*
14	Expenditures, emergency room, out-patient or clinic	1. $1 to 10	1,515	382[e]
		2. 11 to 30	3,385	759
		3. 31 or more	2,798	758
		4. 0	665	114
		Continuous: mean = $44	8,363	2,013[e]
15	Expenditures, family dental	1. $0	6,717	1,444[g]
		2. 1 to 37	4,855	852
		3. 38 to 119	4,885	715
		4. 120 or more	4,743	690
16	Expenditures, family drug	1. $0 to 25	6,921	1,277[g]
		2. 26 to 75	6,918	1,196
		3. 76 or more	7,362	1,228
17	Expenditures, family health	1. $0 to 202	6,725	1,266[g]
		2. 203 to 656	6,930	1,199
		3. 657 or more	7,546	1,236
18	Expenditures, family hospital	1. $0	14,649	2,513[g]
		2. 1 to 386	2,238	415
		3. 387 to 806	2,115	332
		4. 807 or more	2,199	441
19	Expenditures, hospital	1. $0 to 300	1,809	432[r]
		2. 301 to 600	2,263	417
		3. 601 or more	2,355	541
20	Expenditures, hospital, for stay of operation	1. $1 to 300	139	28[m]
		2. 301 to 600	1,610	383
		3. 601 or more	1,736	326
		4. 0, no social survey report	2,038	407

21	Expenditures, out-of-pocket hospital	1. $0	9,002	2,080[l]
		2. 1 to 200	5,299	1,052
		3. 201 or more	860	169
22	Expenditures, out-of-pocket physician visit	1. $0 to 15	33,105	7,504[a]
		2. 16 to 45	10,999	1,914
		3. 46 or more	12,713	2,201
23	Expenditures per admission	1. $0 to 300	2,809	663[c]
		2. 301 to 600	2,776	552
		3. 601 or more	2,437	554
		Continuous: mean = $685	8,022	1,769[c]
24	Expenditures, physician office visits	1. $1 to 15	10,987	1,938[d]
		2. 16 to 45	10,977	1,876
		3. 46 or more	11,041	1,858
		4. 0	1,464	232
		Continuous: mean = $53	34,469	5,904[d]
25	Expenditures, prescribed drugs	1. $0 to 25	46,418	9,611[a]
		2. 26 to 75	5,910	1,107
		3. 76 or more	4,489	901
26	Family income	a. 1. Nonpoor	43,727	6,646[a]
		2. Poor	13,089	4,973
		b. 1. Under $3,000	3,387	1,004[q]
		2. 3,000 to 14,999	12,448	2,520
		3. 15,000 or more	3,289	356
27	Family size	a. 1. 1 or 2 persons	14,582	3,031[a]
		2. 3 to 5	28,777	5,333
		3. 6 persons or more	13,459	3,255
		b. 1. 1 person	4,284	915[q]
		2. 2 persons or more	14,840	2,965
28	Hospital admissions, family	1. 0	14,757	2,540[g]
		2. 1	4,565	775
		3. 2 or more	1,879	386

Table C.1. (Continued)

Variable Number	*Description*	*Classes*	*Weighted N*	*Unweighted N*
29	Hospital admissions per condition	1. 1	11,917	2,515
		2. 2 or more	2,865	692
		3. 0, discovered in verification	380	941
30	Hospital admissions per 100 person-years	Continuous: mean = 143	56,815	11,619[a]
31	Hospital admissions per person	1. 0	50,372	10,228[a]
		2. 1	5,114	1,082
		3. 2 or more	1,331	309
32	Hospital days for stay of operation	1. 1 to 3 days	1,742	343[m]
		2. 4 to 5	937	198
		3. 6 days or more	2,726	578
		4. 0, no social survey report	538	98
33	Hospital days per admission	1. 1 to 4	3,228	678[c]
		2. 5 to 10	3,050	648
		3. 11 or more	1,745	443
		Continuous: mean = 10.4	8,022	1,769[c]
34	Hospital days per condition	1. 1 to 3	2,839	562[l]
		2. 4 to 7	4,316	858
		3. 8 or more	7,626	1,787
		4. 0, discovered in verification	380	94
35	Hospital days per person	1. 1 to 4	2,357	479[r]
		2. 5 to 10	2,288	465
		3. 11 or more	1,781	446
36	Hospital expenditures were reported	1. Yes	4,372	818[a]
		2. No	2,073	573
		3. No hospital stay reported	50,370	10,228
37	Income was reported	1. Yes	43,728	11,228[a]
		2. No	13,090	391

38 to 43	Insurance coverage			
38	Dental coverage	1. Yes	1,910	271[g]
		2. No	18,398	3,255
		3. DK	892	175
39	Drug coverage	1. Yes	1,314	214[g]
		2. No	15,836	2,877
		3. DK	4,050	610
40	Hospital coverage	1. Yes	18,040	3,189[g]
		2. No	2,178	358
		3. DK, NA, found in verification	981	154
41	Major medical coverage	1. Yes	10,250	1,699[g]
		2. No	5,744	980
		3. DK	5,206	1,022
42	Physician visit coverage	1. Yes	2,987	537[g]
		2. No	12,215	2,204
		3. DK	5,998	960
43	Surgical-medical coverage	1. Yes	17,579	3,074[g]
		2. No	2,686	465
		3. DK	935	162
44	Number of calls to complete main questionnaire	a. 1. 1	17,945	4,215[a]
		2. 2	14,219	2,867
		3. 3	8,760	1,754
		4. 4	6,500	1,105
		5. 5 or more	9,360	1,672
		6. DK	31	6
		b. 1. 1	17,945	4,215[a]
		2. 2 to 4	29,479	5,726
		3. 5 or more	9,360	1,672
		4. DK	31	6
45	Number of diagnoses (based on verification data and on social survey when verification unavailable)	1. 0	20,502	4,919[a]
		2. 1	16,003	3,024
		3. 2	10,019	1,772
		4. 3	5,205	970
		5. 4 or more	5,090	934

Table C.1. (Continued)

Variable Number	Description	Classes	Weighted N	Unweighted N
46	Number of hospitalized conditions	a. 1. 0	201	48[i]
		2. 1	3,500	763
		3. 2 or more	1,174	250
		b. 1. 1	3,500	763[i]
		2. 2 to 3	1,171	248
		3. 4 or more	3	2
		4. 0, discovered in verification	201	48
47	Number of physician visit conditions	a. 1. 0	983	197[h]
		2. 1	15,104	2,972
		3. 2 or more	11,838	2,093
		b. 1. 1 to 2	23,022	4,381[h]
		2. 3 to 5	3,853	672
		3. 6 or more	67	12
		4. 0, discovered in verification	983	197
48	Number of surgical procedures (45)	1. 0	286	61[j]
		2. 1	2,261	457
		3. 2 or more	194	46
			Physicians	*Clinics*
49	Number of times verification requested	1. 1	15,233 2,421[n]	1,957 343[p]
		2. 2	8,129 1,408	1,148 224
		3. 3	5,574 867	539 129
		4. 4 to 5	8,822 1,349	1,081 200
		5. 6 to 10	9,692 2,006	990 345
		6. 11 to 20	5,186 1,192	675 376
		7. 21 or more	1,638 505	989 879
50	Operations for hospitalized condition	1. None	8,578	1,937[l]
		2. One or more operations	6,583	1,364

51 to 54	Payments for hospital admissions			
51	Insurance payments	Continuous: mean = $337	8,022	1,769[c]
52	Medicare payments	Continuous: mean = $113	8,022	1,769[c]
53	Out-of-pocket payments	Continuous: mean = $112	8,022	1,769[c]
54	Welfare payments	Continuous: mean = $76	8,022	1,769[c]
55	Perception of health	1. Excellent	21,452	3,910[a]
		2. Good	24,423	5,005
		3. Fair	7,018	1,698
		4. Poor	2,188	590
		5. Deceased, infants, DK, NA	1,734	416
56	Physician Contact	1. Yes	39,104	7,305[a]
		2. No	17,713	4,314
57	Physician cooperation	1. Cooperated	48,206	8,653[n]
		2. Did not cooperate	5,950	1,069
58	Physician expenditures were reported	1. Yes	31,287	2,037[a]
		2. No	7,853	5,292
		3. No physician visit reported	17,675	4,290
59	Physician visits for persons who saw a physician	1. 1	10,438	1,978[b]
		2. 2 to 4	13,674	2,429
		3. 5 or more	14,119	2,725
		Continuous: mean = 5.7	38,437	7,132[b]
60	Physician visits, family	1. 0	1,951	385[g]
		2. 1 to 5	6,276	1,140
		3. 6 to 14	6,753	1,132
		4. 15 or more	6,221	1,044
61	Premium for health insurance policy	1. $1 to 76	4,251	785[g]
		2. 77 to 169	4,239	812
		3. 170 or more	4,192	767
		4. $0, DK, NA	8,519	1,337
		Continuous: mean = $104	19,113	3,280[f]

Table C.1. (Continued)

Variable Number	*Description*	*Classes*	*Weighted N*	*Unweighted N*
62	Race	a. 1. White	49,966	7,855[a]
		2. Nonwhite	6,850	3,764
		b. 1. White	17,055	2,814[q]
		2. Nonwhite	2,070	1,066
63	Regular source of care	1. Yes	37,184	7,592[o]
		2. No	23,718	4,350
64	Residence	a. 1. Rural nonfarm	13,893	2,799[a]
		2. Rural farm	3,897	1,069
		3. SMSA central city	16,909	5,350
		4. SMSA other urban	15,270	1,560
		5. Urban nonSMSA	6,881	841
		b. 1. SMSA	12,737	2,458[q]
		2. NonSMSA	6,387	1,422
65	Response to verification request	1. Person gave permission	49,866	10,583[a]
		2. Person refused permission	6,952	1,036
66	Severity of condition	a. Severity of physician visit condition		
		1. Preventive care	27,503	4,745[k]
		2. Symptomatic relief available	19.942	3,413
		3. Should see doctor	32,626	6,325
		4. Must see doctor	15,235	3,116
		5. DK	192	42
		b. Severity of hospitalized condition		
		1. Preventive care	519	106[l]
		2. Symptomatic relief available	1,678	323
		3. Should see MD	7,425	1,604
		4. Must see MD	5,531	1,264
		5. DK	8	4

67	Severity over all diagnoses (based on verification data and on social survey when verification unavailable)	1. Elective only	13,563	2,349[a]
		2. Elective and mandatory	11,537	2,022
		3. Mandatory only	11,457	2,369
		4. DK, no diagnoses	20,258	4,879
68	Sex	1. Male	27,963	5,515[a]
		2. Female	28,853	6,104
69	Sex and employment status	1. Male, employed full time	18,021	4,025[a]
		2. Male, not employed full time	173	26
		3. Female, employed full time	13,481	2,427
		4. Female, not employed full time	166	24
		5. Unable to work, DK, NA, NAP	24,977	5,117
70	Specialty	1. General practice	25,594	4,880[n]
		2. General surgery	3,448	699
		3. Primary care specialties	12,637	2,063
		4. Other specialties	9,048	1,475
		5. DK, NA, not in the AMA Directory (1969)	3,429	605
71	Survey respondent	a. 1. Self	21,782	4,464[a]
		2. Proxy for child 18 or less	21,230	4,607
		3. Proxy for adult	13,805	2,548
		b. 1. Self	21,782	4,464[a]
		2. Proxy for child 16 or younger	19,253	4,213
		3. Proxy for adult	15,782	2,942
72	Threat level	a. Threat of physician visit conditions		
		1. Very threatening	9,960	1,973[k]
		2. Moderately threatening	12,917	2,427
		3. Nonthreatening	72,260	13,205
		4. DK	361	36
		b. Threat of hospitalized condition		
		1. Very threatening	2,372	512[l]
		2. Moderately threatening	3,009	678
		3. Nonthreatening	9,726	2,098
		4. DK	54	13

Table C.1. (Continued)

Variable Number	*Description*	*Classes*	*Weighted N*	*Unweighted N*
		c. Threat of operation		
		1. Threatening	1,006	183[m]
		2. Nonthreatening	4,519	961
73	Worry about health	1. A great deal	4,762	1,117[a]
		2. Some	10,342	2,057
		3. Hardly any	11,717	2,265
		4. None	28,468	5,824
		5. Infants, DK, NA	1,526	356

Note: Superscripts in "Unweighted N" column denote specific universe being described.

[a]Universe is all sample individuals.
[b]Universe is persons with physician visits reported in social survey.
[c]Universe is admissions reported in social survey.
[d]Universe is persons with office visits reported in social survey.
[e]Universe is persons with emergency room, outpatient, or clinic visits reported in social survey.
[f]Universe is health insurance policies reported in social survey.
[g]Universe is all health insurance policies (reported in social survey or found in verification).
[h]Universe is persons with physician visit conditions (reported in social survey or found in verification).
[i]Universe is persons with hospitalized conditions (reported in social survey or found in verification).
[j]Universe is persons with surgical procedures (reported in social survey or found in verification).
[k]Universe is all social survey and verification physician visit conditions.
[l]Universe is all social survey and verification hospitalized conditions.
[m]Universe is all social survey and verification operations.
[n]Universe is physicians identified as sources of care in social survey or verification (counted each time named by respondents).
[o]Universe is physicians and clinics identified as sources of care in social survey and verification (counted each time named by respondents).
[p]Universe is clinics identified as sources of care in social survey and verification (counted each time named by respondents).
[q]Universe is families.
[r]Universe is persons with admissions reported in the social survey.

References

Allen, J. W. "Interviewer Role Performance: A Further Note on Bias in the Information Interview." *Public Opinion Quarterly,* 1968, *32,* 287–294.

American Hospital Association. *Guide to the Health Care Field.* Chicago: American Hospital Association, 1970.

American Public Health Association. *A Guide to Medical Care Administration.* Vol. 2: *Medical Care Appraisal—Quality and Utilization.* New York: American Public Health Association, 1969.

Andersen, R. "The Effect of Measurement Error on Differences in the Use of Health Services." In R. Andersen, J. Kravits, and O. W. Anderson (Eds.), *Equity in Health Services: Empirical Analyses in Social Policy.* Cambridge, Mass.: Ballinger, 1975.

Andersen, R., Kasper, J., and Frankel, M. R. "The Effect of Measurement Error on Differences in Hospital Expenditures." *Medical Care,* 1976, *14,* 932–949.

Andersen, R., Lion, J., and Anderson, O. W. (Eds.). *Two Decades of Health Services: Social Survey Trends in Use and Expenditure.* Cambridge, Mass.: Ballinger, 1976.

Andersen, R., and others. *Health Service Use: National Trends and Variations: 1953–1971.* DHEW Publication No. (HSM) 73-3004. Washington, D.C.: U.S. Government Printing Office, 1972.

Andersen, R., and others. *Expenditures for Personal Health Services: National Trends and Variations: 1953–1970.* DHEW Publication No. (HRA) 74–3105. Washington, D.C.: U.S. Government Printing Office, 1973.

Bailar, B. "Some Sources of Error and Their Effects on Census Statistics." *Demography,* 1976, *13,* 273–286.

Bailey, L. "Toward a More Complete Analysis of the Total Mean Square Error of Census and Sample Survey Statistics." *Proceedings of the American Statistical Association,* Social Statistics Section, 1975, 1–10.

Bancroft, G., and Welch, E. "Recent Experience with Problems of Labor Force Measurement." *Journal of the American Statistical Association,* 1946, *41,* 303–312.

Belson, W. A. "Respondent Understanding of Survey Questions." *Polls,* 1968, *3* (4), 1–13.

Blue Cross Association. "Report of Survey of Subscriber Understanding of Benefit Booklets." (Mimeograph.) Chicago: Blue Cross Association, 1975.

Cannell, C. F., and Fowler, F. J. "A Study of the Reporting of Visits to Doctors in the National Health Survey." Ann Arbor: Michigan Survey Research Center, University of Michigan, 1963.

Cannell, C. F., and Kahn, R. L. "The Collection of Data by Interviewing." In L. Festinger and D. Katz (Eds.), *Research Methods in the Behavioral Sciences.* New York: Dryden Press, 1953, 327–380.

Cartwright, A. "The Families and Individuals Who Did Not Cooperate on a Sample Survey." *Milbank Memorial Fund Quarterly,* 1959, *37,* 347–368.

Cartwright, A. "Memory Errors in a Morbidity Survey." *Milbank Memorial Fund Quarterly,* 1963, *41,* 5–24.

Cochran, W. G. *Sampling Techniques.* (2nd ed.) New York: Wiley, 1963.

Cochran, W. G. *Sampling Techniques.* (3rd ed.) New York: Wiley, 1978.

Cornfield, J. "On Certain Biases in Samples of Human Populations." *Journal of the American Statistical Association,* 1942, *37,* 63–68.

Dakin, R. E., and Tennant, D. "Consistency of Response by Event-Recall Intervals and Characteristics of Respondents." *Sociological Quarterly,* 1968, *9,* 73–84.

Daniel, W. W. "Nonresponse in Sociological Surveys: A Review of

Some Methods for Handling the Problem." *Sociological Methods and Research,* 1975, *3,* 291–307.

Deming, W. E. "On Errors in Surveys." *American Sociological Review,* 1944, *9,* 359–369.

Deming, W. E. *Some Theory of Sampling.* New York: Wiley, 1950.

Eckler, A. R., and Pritzker, L. "Measuring the Accuracy of Enumerative Surveys." *Bulletin of the International Statistical Institute,* 1951, *33,* 7–24.

Elinson, J., and Trussell, R. E. "Some Factors Relating to Degree of Correspondence for Diagnostic Information as Obtained by Household Interviews and Clinical Examination." *American Journal of Public Health,* 1957, *47,* 311–321.

Feldman, J. J., Hyman, H., and Hart, C. W. "A Field Study of Interviewer Effects on the Quality of Survey Data." *Public Opinion Quarterly,* 1951, *15,* 734–761.

Ferber, R. *The Reliability of Consumer Reports of Financial Assets and Debts.* Studies in Consumer Savings No. 6. Urbana: Bureau of Economic and Business Research, University of Illinois, 1966.

Gray, P. G. "The Memory Factor in Social Surveys." *Journal of the American Statistical Association,* 1955, *50,* 344–363.

Haberman, P. W. "Reliability and Validity of the Data." In J. Kosa and others (Eds.), *Poverty and Health: A Sociological Analysis.* Cambridge, Mass: Harvard University Press, 1969.

Hansen, M. H., Hurwitz, W. N., and Bershad, M. A. "Measurement Errors in Censuses and Surveys." *Bulletin of the International Statistical Institute,* 1961, *29,* 359.

Hansen, M. H., Hurwitz, W. N., and Madow, W. *Sample Survey Methods and Theory.* Vols. 1 and 2. New York: Wiley, 1953.

Hansen, M. H., and Waksberg, J. "Research on Non sampling Errors in Censuses and Surveys." *Review of the International Statistical Institute,* 1970, *38,* 317–332.

Hansen, M. H., and others. "Response Errors in Surveys." *Journal of the American Statistical Association,* 1951, *46,* 147–190.

Health Insurance Institute. *Source Book of Health Insurance Data: 1975–76.* New York: Health Insurance Institute, 1976.

Keyfitz, N. "Estimates of Sampling Variance Where Two Units Are Selected from Each Stratum." *Journal of the American Statistical Association,* 1957, *52,* 503–510.

Kish, L. *Survey Sampling.* New York: Wiley, 1965.

Knudsen, D. D., Pope, H., and Irish, D. P. "Response Differences to Questions on Sexual Standards: An Interview-Questionnaire Comparison." *Public Opinion Quarterly,* 1967, *31,* 290–297.

Koch, G. G. "An Alternative Approach to Multivariate Response Error Models for Sample Survey Data with Applications to Estimators Involving Subclass Means." *Journal of the American Statistical Association,* 1973, *68,* 906–913.

Kosa, J., Alpert, J., and Haggerty, R. "On the Reliability of Family Health Information: A Comparative Study of Mothers' Reports on Illness and Related Behavior." *Social Science and Medicine,* 1967, *1,* 165–181.

Kulley, A. "The Validity of Survey Measurement of Health Services Utilization: A Verification Study of Respondent Reports of Hospitalization." Unpublished doctoral dissertation, Department of Sociology, Purdue University, 1974.

Larson, W. R. "Social Desirability as a Latent Variable in Medical Questionnaire Responses." *Pacific Sociological Review,* 1958, *1* (1), 30–33.

Madow, W. G. "On Some Aspects of Response Error Measurement." *Proceedings of the American Statistical Association,* Social Statistics Section, 1965, 182–192.

Marquis, K. H. "Interviewer-Respondent Interaction in a Household Interview." *Proceedings of the Social Statistical Section of the American Statistical Society,* 1969, 24–30.

Marquis, K. H., Marquis, M. S., and Newhouse, J. P. "The Measurement of Expenditures for Outpatient Physician and Dental Services: Methodological Findings from the Health Insurance Study." *Medical Care,* 1976, *14,* 913–931.

Marquis, K. H., Marshall, J., and Oskamp, S. "Testimony Validity as a Function of Question Form, Atmosphere and Item Difficulty." *Journal of Applied Social Psychology,* 1972, *2,* 167–186.

Mathews, V. L., Feather, J., and Crawford, J. "A Response/Record Discrepancy Study." WHO/ICS-MCU Saskatchewan Study Area Reports, Series 2, No. 2, Saskatoon, Department of Social and Preventive Medicine, University of Saskatchewan, and Springfield, Va., National Technical Information Service PB-226, 327, 1974. (Finalized in 1974 by J. Feather.)

Mauldin, W. P., and Marks, E. S. "Problems of Responses in Enumerative Surveys." *American Sociological Review,* 1950, *15,* 649–657.

Mechanic, D., and Newton, M. "Some Problems in the Analysis of Morbidity Data." *Journal of Chronic Disease,* 1965, *18,* 569–580.

Mettzer, J. W., and Hochstim, J. R. "Reliability and Validity of Survey Data on Physical Health." *Public Health Reports,* 1970, *85,* 1075–1086.

Moser, C. A., and Kalton, G. *Survey Methods in Social Investigation.* New York: Basic Books, 1972.

National Center for Health Services Research. *Advances in Health Survey Research Methods: Proceedings of a National Invitational Conference* (Airlie House, May 1975). DHEW Publication No. (HRA) 77-3154. Washington, D.C.: U.S. Government Printing Office, 1977.

National Center for Health Statistics. *International Classification of Diseases, Adapted for Use in the United States.* (8th rev.) Washington, D.C.: U.S. Public Health Service, 1957.

National Center for Health Statistics.
Series B, No. 32, 1962.
Series 2, No. 6, 1965a.
Series 2, No. 7, 1965b.
Series 2, No. 8, 1965c.
Series 2, No. 18, 1966.
Series 2, No. 23, 1967.
Series 2, No. 26, 1968.
Series 2, No. 36, 1969.
Series 2, No. 43, 1971.
Series 2, No. 48, 1972a.
Series 10, No. 72, 1972b.
Series 2, No. 54, 1973a.
Series 2, No. 57, 1973b.
Series 13, No. 14, 1973c.
Series 10, No. 91, 1974.

National Center for Health Statistics. "Hospitals and Surgical Insurance Coverage Among Persons Under 65 Years of Age in the United States, 1970." *Monthly Vital Statistics Report,* 1972c, *21,* Supplement 2.

Neter J., and Waksberg, J. "A Study of Response Errors in Expenditures Data from Household Interviews." *Journal of the American Statistical Association,* 1964, *59,* 18–55.

Nisselson, H., and Woolsey, T. D. "Some Problems of the Household Interview Design for the National Health Survey." *Journal of the American Statistical Association,* 1959, *54,* 69–87.

Palmer, G. "Factors in the Variability of Response in Enumerative Studies." *Journal of the American Statistical Association,* 1943, *38,* 143–152.

Richardson, A., and Freeman, H. "Evaluation of Medical Care Utilization by Interview Surveys." *Medical Care,* 1972, *10,* 357–362.

Sanders, B. S. "Sources and Validity of Medical Statistics with Special Emphasis on Diagnosis." *Proceeding of the American Statistical Association,* Social Statistics Section, 1963, 250–254.

Social Security Administration. *Compendium of National Health Expenditures Data.* DHEW Publication No. (SSA) 76–11927. Washington, D.C.: U.S. Government Printing Office, 1976.

Solon, J. A., and others. "Patterns of Medical Care: Validity of Interview Information on Use of Hospital Clinics." *Journal of Health and Social Behavior,* 1962, *3,* 21–29.

Stephan, F. F., and McCarthy, P. J. *Sampling Opinions: An Analysis of Survey Procedure.* New York: Wiley, 1958.

Stock, S. J., and Hochstim, J. R. "A Method of Measuring Interviewer Variability." *Public Opinion Quarterly,* 1951, *15,* 322–334.

Suchman, E., and others. "An Analysis of the Validity of Health Questionnaires." *Social Forces,* 1958, *36,* 223–232.

Sudman, S., and Bradburn, N. *Response Effects in Surveys.* Chicago: AVC, 1974.

Sukhatme, P. V. *Sampling Theory of Surveys with Applications.* Ames: Iowa State College Press, 1954.

Thompson, M. M., and Shapiro, G. "The Current Population Survey: An Overview." *Annals of Economic and Social Measurement,* 1973, *2,* 105–129.

United States Bureau of the Census. *The Current Population Survey—Design and Methodology.* Technical Paper No. 40. Washington, D.C.: U.S. Government Printing Office, 1978.

United States Bureau of the Census. Series P-20, No. 225, 1971.

United States Bureau of the Census. Series P-60, No. 80, 1972.

Warwick, D. P., and Lininger, C. A. *The Sample Survey: Theory and Practice.* New York: McGraw-Hill, 1975.

Woolsey, T. D., Lawrence, P., and Balamuth, E. "An Evaluation of Chronic Disease Prevalence Data from HIS." *American Journal of Public Health,* 1962, *52,* 1631–1637.

Zarkovich, S. S. *Sampling Methods and Censuses.* Vol. 2: *Quality of Statistical Data* (Draft). Rome: Food and Agriculture Organization of the United Nations, 1963.

Index

✲ ✲ ✲ ✲ ✲ ✲ ✲ ✲ ✲ ✲ ✲ ✲ ✲ ✲ ✲ ✲ ✲